Erwin Dee Kord (Ed.)

Selway-Bitterroot Wilderness

Erwin Dee Kord (Ed.)

Selway-Bitterroot Wilderness

Strict Nature Reserve/Wilderness Area

Solv

Imprint

Publisher:
Solv is a trademark of
International Book Market Service Ltd., 17 Rue Meldrum, Beau Bassin, 1713-01 Mauritius
Email: info@bookmarketservice.com
Website: www.bookmarketservice.com

Published in 2011

Printed in: U.S.A., U.K., Germany. This book was not produced in Mauritius.

ISBN: 978-613-8-66587-8

Contents

Articles

References

Selway-Bitterroot Wilderness

Selway-Bitterroot Wilderness	
IUCN Category Ib (Wilderness Area)	

Isolated lake in the Selway-Bitterroot Wilderness

Location	Idaho County, Idaho, Ravalli County, Montana, Missoula County, Montana, United States
Nearest city	Darby, Montana
Coordinates	46°12′45″N 114°20′46″W
Area	1,340,460 acres (5,424 km²)

The **Selway-Bitterroot Wilderness** is a protected wilderness area in the states of Idaho and Montana, in the northwestern United States.

At 1.3 million acres (5,300 km²), it is one of the largest designated wilderness areas in the United States (14th overall, but third-largest outside Alaska). It spans the Bitterroot Mountain Range, on the border between Idaho and Montana. It covers parts of Bitterroot National Forest, Clearwater National Forest, Lolo National Forest, and the Nez Perce National Forest. The Frank Church-River of No Return Wilderness Area is immediately to its south, separated only by a dirt road (the Magruder Corridor). Together with adjoining unprotected public land, the two wilderness areas form a five million acre (20,000 km²) wild region.

It is here that the Lochsa and Selway rivers form and flow westward to their confluence at Lowell (outside of the wilderness along U.S. Route 12) to form the Middle Fork of the Clearwater River. The land ranges in elevation from 1700 feet (520 m) on the Lochsa River to 10157 feet (3096 m) at Trapper Peak in the Bitterroot Mountains.

One of the largest elk herds in the United States calls the wilderness home, and it also boasts a healthy population of bighorn sheep. This wilderness is one of the areas being used to reintroduce wild packs of the grey wolf.

The high-elevation areas around the Bitterroot peaks are rugged alpine parkland. Lower areas are covered by dense coniferous forest.

See also

- List of largest wilderness areas in the United States
- List of U.S. Wilderness Areas
- Wilderness Act

External links

- Selway-Bitterroot Wilderness Information & Maps [1]
- Selway-Bitterroot Wilderness [2]—Information from the Bitterroot National Forest
- Wilderness.net [3]
- TopoQuest [4]
- National Public Radio [5] report on the Wilderness Act from the Selway-Bitterroot Wilderness]

References

[1] http://explorethebitterroot.com/selway-bitterroot-wilderness
[2] http://www.fs.fed.us/r1/bitterroot/recreation/wilderness/selway_bitterroot.htm
[3] http://www.wilderness.net/index.cfm?fuse=NWPS&sec=wildView&wname=Selway-Bitterroot
[4] http://www.topoquest.com/map.asp?z=11&n=5132020&e=697581&s=50&size=l&u=7&datum=nad83&layer=DRG25
[5] http://www.npr.org/programs/re/archivesdate/2004/sep/selway/

Strict Nature Reserve/Wilderness Area

A **Strict Nature Reserve** (**IUCN Ia**) or **Wilderness Area** (**IUCN Ib**), is the highest category of protected area recognised by the World Commission on Protected Areas (WCPA), a body which is part of the International Union for Conservation of Nature and Natural Resources (IUCN). These category I areas are the most stringently protected natural landscapes.

Banner Peak above Thousand Island Lake in a US Wilderness Area

Purpose

Strict Nature Reserves and Wilderness Areas are protected areas that are created and managed mainly for the purposes of research or for the protection of large, unspoilt areas of wilderness. Their primary purpose is the preservation of biodiversity and as essential reference areas for scientific work and environmental monitoring.

- IUCN Ia *Strict Nature Reserves* are generally established exclusively for scientific field work.
- IUCN Ib *Wilderness Areas* are defined as "large unmodified or slightly modified areas, retaining their natural character and influence, without permanent or significant human habitation, which are protected and managed so as to preserve their natural condition."

Usage and intrusion are strictly controlled. As a result Strict Nature Reserves often form the core zones, with Wilderness Areas acting as a buffer zone, similar to the concept used for national parks (which are IUCN Category II), but also for UNESCO World Heritage Sites.

Protected areas

Austria

In Austria, only one area has been designated as a IUCN Category I protected area:

- *Urwald Rothwald* (IUCN Ia) in the *Wildnisgebiet Dürrenstein* (Ib), Lower Austrian *Eisenwurzen*

Germany

In their national strategy for biological diversity out to 2020 published in 2007, the German Federal Government set the aim of allowing 2% of the area of Germany to develop naturally and in an undisturbed way into wilderness areas.[1]

In Germany the following IUCN Category I area has been designated:

- *Königsbrücker Heide* (IUCN Ib)

United States

In the USA *Wilderness Areas* are the strictest category of protected area. They are designated by law in accordance with the Wilderness Act of 1964 by the United States Congress. They must cover an area of at least 20 km² (with the exception of islands) and may only be entered on foot, by canoe or on horseback. Exceptions are the extensive areas of wilderness in Alaska. At the beginning of 2010 there were over 750 recognised Wilderness Areas.

See also

- Wilderness
- IUCN Protected Area Management

Sources

- Nigel Dudley (Ed.): *Guidelines for applying protected area management categories.* IUCN, Gland , ISBN 978-2-8317-1086-0 (Weblink [2], download pdf, iucn.org).

References

[1] www.blickpunkt-brandenburg.de. (http://www.blickpunkt-brandenburg.de/jueterbogluckenwalde/nachrichten/?doc=10687) Accessed on 11 Jun 2011.

[2] http://www.iucn.org/about/union/commissions/wcpa/wcpa_puball/wcpa_pubsubject/wcpa_categoriespub/?1662/Guidelines-for-applying-protected-area-management-categories

External links

- IUCN - World Commission on Protected Areas (http://cms.iucn.org/about/union/commissions/wcpa/index.cfm) (engl.)
- IUCN Protected Area Management Guidelines (http://www.wild.org/main/policy-research/iucn-guidelines-for-applying-protected-area-management-catagories/), The Wild Foundation (engl.)
- IUCN-Schutzgebietskategorien (pdf) (http://www.wwf.de/fileadmin/fm-wwf/pdf_neu/IUCN_Schutzgebietskriterien.pdf)

Bestand:

- World Database of Protected Areas (http://www.unep-wcmc.org/wdpa/index.htm), unep-wcmc.org

Idaho

State of Idaho	
 Flag	 Seal
Nickname(s): *Gem State*	
Motto(s): *Esto perpetua*	
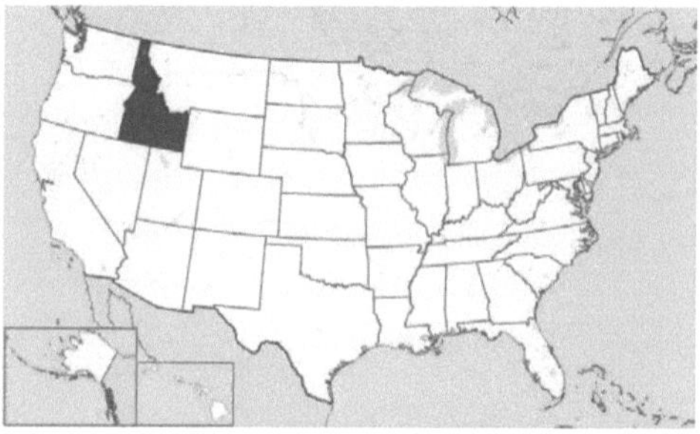	
Official language(s)	English
Demonym	Idahoan
Capital (and largest city)	Boise
Largest metro area	Boise metropolitan area
Area	Ranked 14th in the U.S.
- Total	83,570 sq mi (216,632 km^2)
- Width	305 miles (491 km)
- Length	479 miles (771 km)
- % water	0.98
- Latitude	42° N to 49° N
- Longitude	111°03′ W to 117°15′ W
Population	Ranked 39th in the U.S.
- Total	1,567,582 (2010 Census)
- Density	15.64/sq mi (6.04/km^2) Ranked 44th in the U.S.
Elevation	

- Highest point	Borah Peak[1] [2] [3] 12,668 ft (3861.2 m)
- Mean	5,000 ft (1,520 m)
- Lowest point	Confluence of Snake River and Clearwater River[2] [3] 713 ft (217 m)
Admission to Union	July 3, 1890 (43rd)
Governor	C.L. "Butch" Otter (R)
Lieutenant Governor	Brad Little (R)
Legislature	Idaho Legislature
- Upper house	Senate
- Lower house	House of Representatives
U.S. Senators	Mike Crapo (R) Jim Risch (R)
U.S. House delegation	1-Raúl Labrador (R) 2-Mike Simpson (R) (list)
Time zones	
- north of Salmon River	Pacific: UTC−8/−7
- remainder	Mountain: UTC−7/−6
Abbreviations	ID US-ID
Website	www.idaho.gov [4]

Idaho i/ˈaɪdəhoʊ/ is a state in the Rocky Mountain area of the United States. The state's largest city and capital is Boise. Residents are called "Idahoans". Idaho was admitted to the Union on July 3, 1890, as the 43rd state.

Idaho is a mountainous state, and is the 14th largest state by land area, with an area larger than that of all of New England. It is landlocked, surrounded by the states of Washington, Oregon, Nevada, Utah, Wyoming, Montana and the Canadian province of British Columbia. However, the network of dams and locks on the Columbia River and Snake River make the city of Lewiston the farthest inland seaport on the Pacific coast of the contiguous United States.

According to the United States Census 2010, Idaho had a population of 1,567,582. The state's postal abbreviation is **ID**. Idaho's nickname is the **Gem State**, because nearly every known gemstone has been found there.[5] In addition, Idaho is one of only two places in the world where star garnets can be found (the other is the Himalaya Mountains, in Pakistan and India), and is the only place six-pointed star garnets have been found. Additionally Idaho is sometimes called the **Potato State** owing to its popular crop. The state motto is *Esto Perpetua* (Latin for "Let it be forever").

Etymology

The exact origin of the name remains a mystery.[6] In the early 1860s, when the United States Congress was considering organizing a new territory in the Rocky Mountains, eccentric lobbyist George M. Willing suggested the name "Idaho," which he claimed was derived from a Shoshone language term meaning "the sun comes from the mountains" or "gem of the mountains". Willing later claimed that he had made up the name himself.[7] [8] Congress ultimately decided to name the area Colorado Territory when it was created in February 1861. Thinking they would get a jump on the name, locals named a community in Colorado "Idaho Springs".

However, the name "Idaho" did not go away. The same year Congress created Colorado Territory, a county called Idaho County was created in eastern Washington Territory. The county was named after a steamship named Idaho, which was launched on the Columbia River in 1860. It is unclear whether the steamship was named before or after Willing's claim was revealed. Regardless, a portion of Washington Territory, including Idaho County, was used to create Idaho Territory in 1863.

Lake Coeur d'Alene in North Idaho

Despite this lack of evidence for the origin of the name, many textbooks well into the 20th century repeated as fact Willing's account that the name "Idaho" derived from the Shoshone term "ee-da-how".

The name "Idaho" may be derived from the Plains Apache word "ídaahę" which means "enemy." The Comanches used this word to refer to the Idaho Territory.[9]

An excerpt from a 1956 Idaho History Textbook:

> "Idaho" is a Shoshoni Indian exclamation. The word consists of three parts. The first is "Ee", which in English conveys the idea of "coming down". The second is "dah" which is the Shoshoni stem or root for both "sun" and "mountain". The third syllable, "how", denotes the exclamation and stands for the same thing in Shoshoni that the exclamation mark (!) does in the English language. The Shoshoni word is "Ee-dah-how", and the Indian thought thus conveyed when translated into English means, "Behold! the sun coming down the mountain".[10]

Geography

Further information: List of Idaho counties

Idaho borders six states and one Canadian province. The states of Washington and Oregon are to the west, Nevada and Utah are to the south, and Montana and Wyoming are to the east. Idaho also shares a short border with the Canadian province of British Columbia to the north. The landscape is rugged with some of the largest unspoiled natural areas in the United States. For example, at 2.3 million acres (9,300 km²), the Frank Church-River of No Return Wilderness Area is the largest contiguous area of protected wilderness in the continental United States. Idaho is a Rocky Mountain state with abundant natural resources and scenic areas. The state has snow-capped mountain ranges, rapids, vast lakes and steep canyons. The waters of the Snake River rush through Hells Canyon, the deepest gorge in the United States.

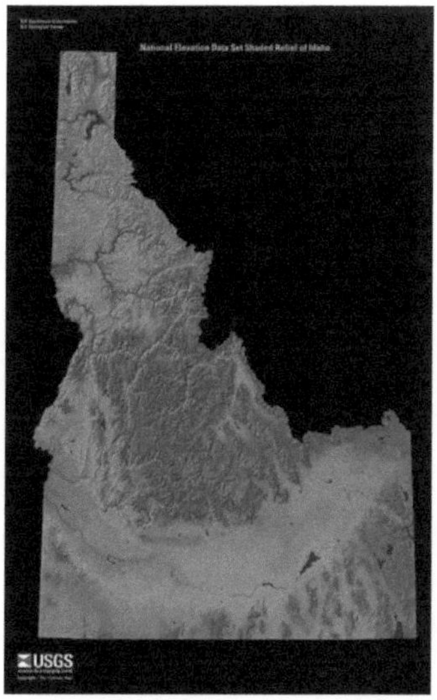

Digitally colored elevation map of Idaho.

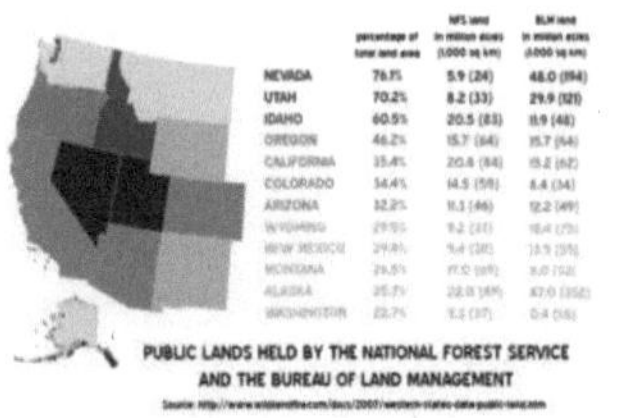

Sixty percent of Idaho's land is held by the National Forest Service or the Bureau of Land Management, and it leads the nation in forest service land as a percentage of total area.[11] [12]

Reynolds Creek Experimental Watershed in the Owyhee Mountains about 50 miles (80 km) southwest of Boise, Idaho.

Redfish Lake in central Idaho.

The Palouse region of North Idaho.

The Shoshone Falls in southern Idaho.

Shoshone Falls plunges down rugged cliffs from a height greater than that of Niagara Falls. The major rivers in Idaho are the Snake River, the Clark Fork/Pend Oreille River, the Clearwater River, the Salmon River. Other significant rivers include the Coeur d'Alene River, the Spokane River, the Boise River, and the Payette River. The Salmon River empties into the Snake in Hells Canyon and forms the southern boundary of Nez Perce County on its north shore, of which Lewiston is the county seat. The Port of Lewiston, at the confluence of the Clearwater and the Snake Rivers is the farthest inland seaport on the West Coast at 465 river miles from the Pacific at Astoria, Oregon.[13]

Idaho's highest point is Borah Peak, 12662 ft (3859 m), in the Lost River Range north of Mackay. Idaho's lowest point, 710 ft (216 m), is in Lewiston, where the Clearwater River joins the Snake River and continues into Washington. The Sawtooth Range is often considered Idaho's most famous mountain range.[14] Other mountain ranges in Idaho include the Bitterroot Range, the White Cloud Mountains, the Lost River Range, the Clearwater Mountains, and the Salmon River Mountains.

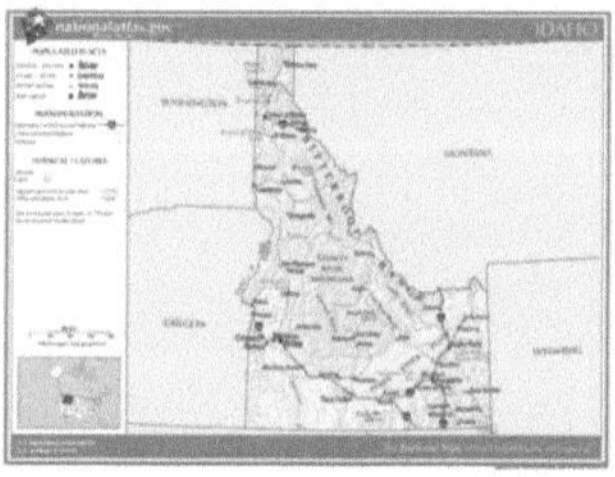

Map of Idaho

Southern Idaho, including the Boise metropolitan area, Idaho Falls, Pocatello, and Twin Falls are in the Mountain Time Zone. (A legislative oddity (15 U.S.C. ch.6 [15] §264) theoretically placed this region in the Central Time Zone, but this error was corrected with a 2007 Amendment.)[16] Areas north of the Salmon River, including Coeur d'Alene, Moscow, Lewiston, and Sandpoint are in the Pacific Time Zone and revolve commercially and culturally around Washington State through the second largest city, Spokane, Washington.

Climate

Idaho has much variation in its climate. Although the state's western border is located about 350 miles (560 km) from the Pacific Ocean, the maritime influence is still felt in Idaho, especially in the winter when cloud cover, humidity, and precipitation are at their maximum extent. This influence has a moderating effect in the winter where temperatures are not as low as would otherwise be expected for a northern state with a predominantly elevated altitude.[17] The maritime influence is the least prominement in the eastern part of the state where the precipitation patterns are often reversed, with wetter summers and drier winters, and seasonal temperature differences more extreme, showing a more semi-arid continental climate.

Climate in Idaho can be hot, although extended periods over 100 °F (38 °C) for the maximum temperature are rare, except for the lowest point in elevation, Lewiston, which correspondingly sees very little snow. Hot summer days are tempered by the low relative humidity and cooler evenings during summer months since, for most of the state, the highest diurnal difference in temperature is often in the summer. Winters can be cold, although extended periods of bitter cold weather below zero are unusual. This is what led the railroad tycoon Harriman family to develop the most famous ski resort, Sun Valley. Idaho's all time highest temperature of 118 °F (48 °C) was recorded at Orofino on July 28, 1934; the all time lowest temperature of −60 °F (−51 °C) was recorded at Island Park Dam on January 18, 1943.

Monthly Normal High and Low Temperatures For Various Idaho Cities. (F)

City	Jan	Feb	Mar	Apr	May	Jun	Jul	Aug	Sep	Oct	Nov	Dec
Boise	38/24	45/27	55/33	62/38	72/46	81/53	91/59	90/59	79/50	65/40	48/31	38/23
Lewiston	42/30	47/31	55/36	62/41	71/47	79/54	89/60	89/60	78/51	63/41	48/34	40/28
Pocatello	33/16	38/19	49/27	59/33	68/40	78/46	88/52	88/51	76/42	62/33	45/24	33/16
[18]												

Lakes

- Alturas Lake
- Bear Lake (Idaho-Utah)
- Dirkies Lake
- Hayden Lake
- Henry's Lake
- Hidden Lake
- Lake Cascade
- Lake Coeur d'Alene
- Lake Lowell
- Lake Ruvet
- Lake Walcott
- Payette Lake (McCall)
- Pend Oreille
- Little Redfish Lake
- Lucky Peak Lake
- Pettit Lake
- Priest Lake
- Redfish Lake
- Sawtooth Lake
- Stanley Lake
- Warm Lake
- Williams Lake (Salmon)

History

Humans may have been present in the Idaho area as long as 14,500 years ago. Excavations at Wilson Butte Cave near Twin Falls in 1959 revealed evidence of human activity, including arrowheads, that rank among the oldest dated artifacts in North America. American Indian peoples predominant in the area included the Nez Perce in the north and the Northern and Western Shoshone in the south.

An early presence of French-Canadian trappers is visible in names and toponyms that have survived to this day: Nez Percé, Coeur d'Alène, Boisé, Payette, some preexisting the Lewis and Clark and Astorian expeditions which themselves included significant numbers of French and Metis guides recruited for their familiarity with the terrain.

Idaho, as part of the Oregon Country, was claimed by both the United States and Great Britain until the United States gained undisputed jurisdiction in 1846. From 1843 to 1849 present-day Idaho was under the *de facto* jurisdiction of the Provisional Government of Oregon. When Oregon became a state, what is now Idaho was in what was left of the original Oregon Territory not part of the new state, and designated as the Washington Territory.

Between then and the creation of the Idaho Territory on July 4, 1863 at Lewiston, parts of the present-day state were included in the Oregon, Washington, and Dakota Territories. The new territory included present-day Idaho, Montana, and most of Wyoming. The Lewis and Clark expedition crossed Idaho in 1805 on the way to the Pacific and in 1806 on the return, largely following the Clearwater River both directions. The first non-indigenous settlement was Kullyspell House, established on the shore of Lake Pend Oreille for fur trading in 1809 by David Thompson of the North West Company.[19] [20] In 1812 Donald Mackenzie, working for the Pacific Fur Company at the time, established a post on the lower Clearwater River near present-day Lewiston. This post, known as "MacKenzie's Post" or "Clearwater", operated until the Pacific Fur Company was bought out by the North West Company in 1813, after which it was abandoned.[21] [22] The first attempts at organized communities, within the present borders of Idaho, were established in 1860.[23] [24] The first permanent, substantial incorporated community was Lewiston in 1861.

After some tribulation as a territory, including the illegal and chaotic transfer of the territorial capital from Lewiston in December 1864 to Boise in January 1865, disenfranchisement of Mormon polygamists upheld by the U.S. Supreme Court in 1877,[25] and a federal attempt to split the territory between Washington Territory which gained statehood in 1889, a year before Idaho, and the state of Nevada which had been a state since 1863, Idaho achieved statehood in 1890. The economy of the state, which had been primarily supported by metal mining, shifted towards agriculture, forest products and tourism.

In recent years, Idaho has expanded its commercial base as a tourism and agricultural state to include science and technology industries. Science and technology have become the largest single economic center (over 25% of the state's total revenue) within the state and are greater than agriculture, forestry and mining combined.[26]

The Idaho State Historical Society and numerous local historical societies and museums preserve and promote Idaho's cultural heritage.

Demographics

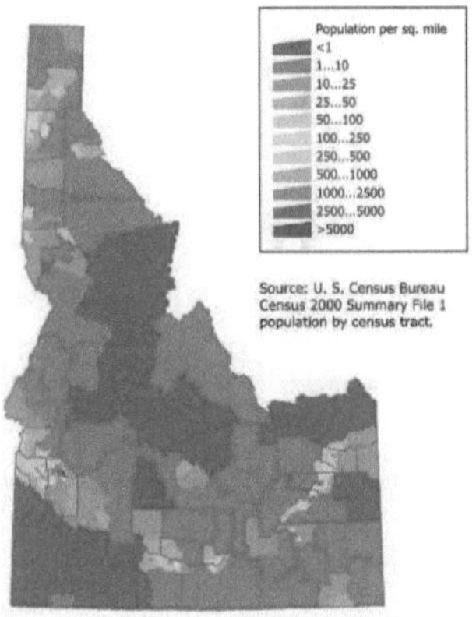

Idaho Population Density Map.

Historical populations		
Census	**Pop.**	**%±**
1870	14999	—
1880	32610	117.4%
1890	88548	171.5%
1900	161772	82.7%
1910	325594	101.3%
1920	431866	32.6%
1930	445032	3.0%
1940	524873	17.9%
1950	588637	12.1%
1960	667191	13.3%
1970	712567	6.8%
1980	943935	32.5%
1990	1006749	6.7%
2000	1293953	28.5%
2010	1567582	21.1%
Source: 1910–2010[27]		

At the 2010 Census, 84.0% of the population was non-Hispanic White, 0.6% non-Hispanic Black or African American, 1.1% non-Hispanic American Indian and Alaska Native, 1.2% non-Hispanic Asian, 0.1% non-Hispanic Native Hawaiian and Other Pacific Islander, 0.1% from some other race (non-Hispanic) and 1.7% of two or more races (non-Hispanic). 11.2% of Idaho's population was of Hispanic, Latino, or Spanish origin (they may be of any race).

As of 2005, Idaho has an estimated population of 1,429,096, which is an increase of 33,956, or 2.4%, from the prior year and an increase of 135,140, or 10.4%, since the year 2000. This includes a natural increase since the last census of 58,884 people (that is 111,131 births minus 52,247 deaths) and an increase due to net migration of 75,795 people into the state. There are large numbers of Americans of English and German ancestry in Idaho. Immigration from outside the United States resulted in a net increase of 14,522 people, and migration within the country produced a net increase of 61,273 people.

This made Idaho the sixth fastest-growing state after Arizona, Nevada, Florida, Georgia, and Utah. From 2004 to 2005, Idaho grew the third-fastest, surpassed only by Nevada and Arizona.

Nampa, the state's second largest city, has experienced particularly strong growth in recent years. According to census estimates Nampa has grown 22.1% to nearly 65,000 residents between 2000 and 2003. As of 2007, the population in Nampa was estimated at 84,000. Growth of 5% or more over the same period has also been observed in Caldwell, Coeur d'Alene, Meridian, Post Falls and Twin Falls.[28]

Since 1990, Idaho's population has increased by 386,000 (38%).

The Boise Metropolitan Area (officially known as the Boise City-Nampa, ID Metropolitan Statistical Area) is Idaho's largest metropolitan area. Other metropolitan areas in order of size are Coeur d'Alene, Idaho Falls, Pocatello and Lewiston.

As of 2006, six official micropolitan statistical areas are based in Idaho. Twin Falls is the largest of these.

The center of population of Idaho is located in Custer County, in the town of Stanley.[29]

The most common reported ancestries in the state are: German (18.9%), English (18.1%), Irish (10%), American (8.4%), Norwegian (3.6%), and Swedish (3.5%).

Religion

According to a report produced by the Pew Forum on Religion & Public Life the self-identified religious affiliations of Idahoans over the age of 18 as of 2008 are:[30]

The Church of Jesus Christ of Latter-day Saints Idaho Falls Temple.

Denomination	%
Church of Jesus Christ of Latter-day Saints	23
Evangelical Protestant	22
Catholic	18
Unaffiliated	18
Mainline Protestant	16
Other Faiths	2
Jehovah's Witness	1
Eastern Orthodox	<0.5
Other Christian	<0.5
Jewish	<0.5
Muslim	<0.5
Buddhist	<0.5
Hindu	<0.5
Other World	<0.5
Don't know/refused	<0.5

The largest denominations by number of members in 2000 were The Church of Jesus Christ of Latter-day Saints with 311,425; (2008: 406,764) the Roman Catholic Church with 130,847; the Assemblies of God with 18,745; and the United Methodist Church with 17,683.[31]

Economy

Idaho State Quarter

American Falls Dam

Gross state product for 2004 was US$43.6 billion. The per capita income for 2004 was US$26,881. Idaho is an important agricultural state, producing nearly one-third of the potatoes grown in the United States. All three varieties of wheat, Dark Northern Spring, Hard Red and Soft White are grown in the state. Nez Perce County is considered a premier Soft White growing locale.

Important industries in Idaho are food processing, lumber and wood products, machinery, chemical products, paper products, electronics manufacturing, silver and other mining, and tourism. The world's largest factory for barrel cheese, the raw product for processed cheese is located in Gooding, Idaho. It has a capacity of 120,000 metric tons per year of barrel cheese and belongs to the Glanbia group.[32] The Idaho National Laboratory (INL), a government lab for nuclear energy research, is also an important part of the eastern Idaho economy. Idaho also is home to three facilities of Anheuser-Busch which provide a large part of the malt for breweries located across the nation.

Locally, a variety of industries are important. Outdoor recreation is a common example ranging from numerous snowmobile and downhill and cross-country ski areas in winter to the evolution of Lewiston as a retirement community based on mild winters, dry year-round climate and one of the lowest median wind velocities anywhere, combined with the rivers for a wide variety of activities. Other examples would be ATK Corporation, which operates three ammunition and ammunition components plants in Lewiston. Two are sporting and one is defense contract. The Lewis-Clark Valley has an additional independent ammunition components manufacturer and the Chipmunk rifle factory. Four of the world's six welded aluminum jet boat (for running river rapids) manufacturers are in the Lewiston-Clarkston, WA valley. Wine grapes were grown between Kendrick and Julietta in the Idaho Panhandle by the French Rothchilds until Prohibition. In keeping with this, while there are no large wineries or breweries in Idaho, there are numerous and growing numbers of award winning boutique wineries and microbreweries in the northern part of the state.

Today, the largest industry in Idaho is the science and technology sector. It accounts for over 25% of the State's total revenue and 70%+ of the State's exports (in dollars). Idaho's industrial economy is growing, with high-tech products leading the way. Since the late 1970s, Boise has emerged as a center for semiconductor manufacturing. Boise is the home of Micron Technology Inc., the only U.S. manufacturer of dynamic random access memory (DRAM) chips. Micron at one time manufactured desktop computers, but with very limited success. Hewlett-Packard has operated a large plant in Boise since the 1970s, which is devoted primarily to LaserJet printers production.[33] ON Semiconductor, whose worldwide headquarters in Pocatello, is a widely recognized innovator in modern integrated mixed-signal semiconductor products, mixed-signal foundry services, and structured digital products. Coldwater Creek, a women's clothing retailer, is headquartered in Sandpoint. Sun Microsystems (now a part of Oracle Corporation) has two offices in Boise and a parts depot in Pocatello. Sun brings $4M in annual salaries and over $300M of revenue to the state each year.

A number of Fortune 500 companies started in or trace their roots to Idaho, including JC Penney (as The Golden Rule) in Twin Falls, Safeways in American Falls, Albertsons in Boise, JR Simplot across southern Idaho, Potlatch Corp. in Lewiston and Zimmerly Air Transport in Lewiston-Clarkston was one of the five companies in the merger centered around Varney Flying Service of Pasco, Washington, which became United Air Lines and subsequently Varney Air Group that became Continental Airlines.

Wheat harvest on the Palouse

The state personal income tax ranges from 1.6% to 7.8% in eight income brackets. Idahoans may apply for state tax credits for taxes paid to other states, as well as for donations to Idaho state educational entities and some nonprofit youth and rehabilitation facilities.

The state sales tax is 6% with a very limited, selective local option up to 6.5%. Sales tax applies to the sale, rental or lease of tangible personal property and some services. Food is taxed, but prescription drugs are not. Hotel, motel, and campground accommodations are taxed at a higher rate (7% to 11%). Some jurisdictions impose local option sales tax.

Idaho has a state gambling lottery which contributed $333.5 million in payments to all Idaho public schools and Idaho higher education from 1990–2006.[34]

Energy

The energy landscape of Idaho is favorable to the development of renewable energy systems. The state is rich in renewable energy resources but has limited fossil fuel resources. The Snake River Plain and smaller river basins provide Idaho with some of the best hydroelectric power resources in the nation and its geologically active mountain areas have significant geothermal power and wind power potential. These realities have shaped much of the state's current energy landscape.

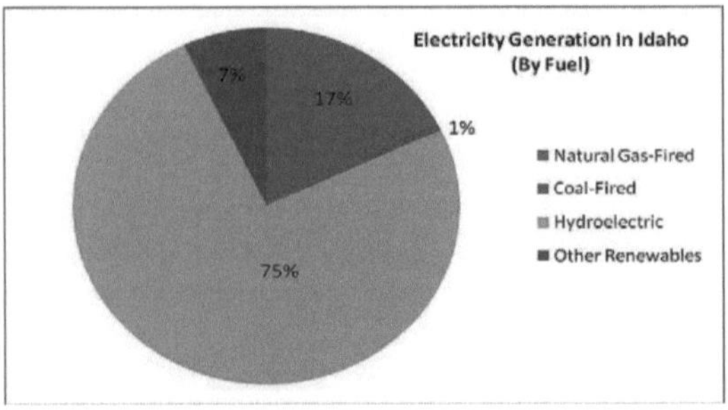

Electricity Generation in Idaho

Most of the energy consumed in Idaho is imported from other states. Imports account for more than 80% of total energy consumption, including all of Idaho's natural gas and petroleum supplies and more than half of its electricity. Of the electricity consumed in Idaho in 2005, 48% came from hydroelectricity, 42% was generated by burning coal and 9% was generated by burning natural gas. The remainder came from other renewable sources such as wind.[35]

The state's numerous river basins allow hydroelectric power plants to provide 556 thousand MWh, which amounts to about three-fourths of Idaho's electricity generated in the state. Washington State provides most of the natural gas used in Idaho through one of the two major pipeline systems supplying the state. Although the state relies on out-of-state sources for its entire natural gas supply, it uses natural gas-fired plants to generate 127 thousand MWh, or about ten percent of its output. Coal-fired generation and the state's small array of wind turbines supplies the remainder of the state's electricity output. The state produces 739 thousand MWh but still needs to import half of its electricity from out-of-state to meet demand.[36]

While Idaho's 515 trillion Btu total energy consumption is low compared with other states and represents just 0.5% of United States consumption, the state also has the nation's 11th smallest population, 1.5 million, so its per capita energy consumption of 352 million Btu is currently just above the national average of 333 million Btu.[36] As the 13th largest state in land area, distance creates the additional problem of "line loss". When the length of an electrical transmission line is doubled, the resistance to an electric current passing through it is also doubled.

In addition, Idaho also has the 6th fastest growing population in the United States with the population expected to increase by 31% from 2008 to 2030.[37] This projected increase in population will contribute to a 42% increase in demand by 2030, further straining Idaho's finite hydroelectric resources.[38]

Transportation

Major highways

The current state license plate design, modified since its introduction in 1991.

Idaho is among the few states in the nation without a major freeway linking the two largest metropolitan areas of Boise in the south and Coeur d'Alene in the north. US 95 links the two ends of the state, but like many other highways in Idaho, it is badly in need of repair and upgrade. In 2007, the Idaho Transportation Department stated that the state's highway infrastructure faces a $200 million per year shortfall in maintenance and upgrades. Interstate 84 is the main highway linking the Southeast and Southwest portions of the state, along with Interstate 86 and Interstate 15.

Interstate 15.

Major federal aid highways in Idaho:

U.S. Route 95.

North	North/South	West/East	Southwest
• U.S. Highway 2	• U.S. Highway 93	• U.S. Highway 20	• Interstate 184
• U.S. Highway 12	• U.S. Highway 95	• U.S. Highway 26	
	• Interstate 15	• U.S. Highway 30	
		• Interstate 84	
		• Interstate 86	
		• Interstate 90	

Air Travel

Major airports include the Boise Airport serving the southwest region of Idaho, and the Spokane International Airport (located in Spokane, Washington), which serves northern Idaho. Other airports with scheduled service are the Pullman-Moscow Regional Airport serving the Palouse; the Lewiston-Nez Perce County Airport, serving the Lewis-Clark Valley and north central and west central Idaho; The Magic Valley Regional Airport in Twin Falls; the Idaho Falls Regional Airport; and the Pocatello Regional Airport.

Rail Travel

Idaho is served by two transcontinental railroads. The Burlington Northern Santa Fe (BNSF) connects North Idaho with Seattle, Portland and Spokane to the west, and Minneapolis and Chicago to the east. The BNSF travels through Kootenai, Bonner and Boundary Counties. The Union Pacific Railroad crosses southern Idaho traveling between Portland, Green River, WY, and Ogden, Utah and serves Boise, Nampa, Twin Falls, and Pocatello. Amtrak's Empire Builder crosses northern Idaho, with its only stop being in Sandpoint. There has been a push recently to return Amtrak service to southern Idaho, as well.

Ports

The Port of Lewiston is the farthest inland Pacific port on the west coast. A series of dams and locks on the Snake River and Columbia River facilitate barge travel from here to Portland, where goods are loaded on ocean-going vessels.

Law and government

State Constitution

The Constitution of Idaho is roughly modeled on the national constitution with several additions. The constitution defines the form and functions of the state government, and may be amended through plebiscite. Notably, the state constitution presently requires the state government to maintain a balanced budget. As result, Idaho has limited debt (construction bonds, etc.).

The Idaho State Capitol building in Boise.

Idaho Code

All of Idaho's state laws are contained in the Idaho Code. The code is amended through the Legislature with the approval of the Governor. Idaho still follows its original state Constitution.

State government

The constitution of Idaho provides for three branches of government: the executive, legislative and judicial branches. Idaho has a bicameral legislature, elected from 35 legislative districts, each represented by one senator and two representatives. Idaho still operates under its original (1889) state constitution.

Since 1946, statewide elected constitutional officers have been elected to four-year terms. They include: Governor, Lieutenant Governor, Secretary of State, Controller (Auditor before 1994), Treasurer, Attorney General, and Superintendent of Public Instruction.

Last contested in 1966, Inspector of Mines was an original elected constitutional office. Afterward it was an appointed position and ultimately done away with entirely in 1974.

Idaho's government has an alcohol monopoly.

Executive Branch

Further information: List of Idaho Governors

Further information: Lieutenant Governor of Idaho

Further information: Secretary of State of Idaho

The governor of Idaho serves a four-year term, and is elected during what is nationally referred to as midterm elections. As such, the governor is not elected in the same election year as the president of the United States. The current governor is Republican C. L. "Butch" Otter, who was elected in 2006 and re-elected in 2010.

Legislative Branch

Idaho's legislature is part-time. However, the session may be extended if necessary, and often is. Because of this, Idaho's legislators are considered "citizen legislators", meaning that their position as a legislator is not their main occupation.

Terms for both the Senate and House of Representatives are two years. Legislative elections occur every even numbered year.

The Idaho Legislature has been continuously controlled by the Republican Party since the late 1950s, although Democratic legislators are routinely elected from Boise, Pocatello, Blaine County and the northern Panhandle.

See also List of Idaho senators and representatives

Judicial Branch

The highest court in Idaho is the Idaho Supreme Court. There is also an intermediate appellate court, the Idaho Court of Appeals, which hears cases assigned to it from the Supreme Court. The state's District Courts serdistricts.[39]

Counties

Idaho is divided into political jurisdictions designated as *counties*. Since 1919 there are 44 counties in the state, ranging in size from 410 to 8,502 square miles (1,062 to 22,020 square kilometers).

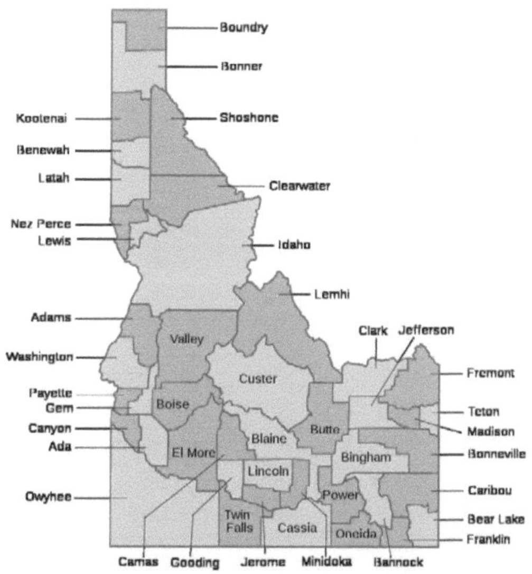

County name	County seat	Year founded	Population 2008 Est.	Population Percentage	Area (sq. m.)	Area Percentage
Ada	Boise	1864	380,920	25.00 %	1,060	1.21 %
Adams	Council	1911	3,499	00.23 %	1,370	1.57 %
Bannock	Pocatello	1893	80,812	05.30 %	1,147	1.31 %
Bear Lake	Paris	1893	5,798	00.38 %	1,049	1.20 %
Benewah	St. Maries	1915	9,352	00.61 %	784	0.90 %
Bingham	Blackfoot	1885	43,903	02.88 %	2,120	2.42 %
Blaine	Hailey	1895	21,731	01.43 %	2,661	3.04 %
Boise	Idaho City	1864	7,504	00.49 %	1,907	2.18 %
Bonner	Sandpoint	1907	41,168	02.70 %	1,920	2.19 %
Bonneville	Idaho Falls	1911	99,135	06.51 %	1,901	2.17 %
Boundary	Bonners Ferry	1915	10,962	00.72 %	1,278	1.46 %
Butte	Arco	1917	2,751	00.18 %	2,234	2.55 %
Camas	Fairfield	1917	1,126	00.07 %	1,079	1.23 %
Canyon	Caldwell	1891	183,939	12.07 %	604	0.69 %
Caribou	Soda Springs	1919	6,826	00.45 %	1,799	2.06 %
Cassia	Burley	1879	21,348	01.40 %	2,580	2.95 %
Clark	Dubois	1919	910	00.06 %	1,765	2.02 %
Clearwater	Orofino	1911	8,176	00.54 %	2,488	2.84 %
Custer	Challis	1881	4,254	00.28 %	4,937	5.64 %
Elmore	Mountain Home	1889	28,997	01.90 %	3,101	3.54 %
Franklin	Preston	1913	12,454	00.82 %	668	0.76 %
Fremont	St. Anthony	1893	12,551	00.82 %	1,896	2.17 %
Gem	Emmett	1915	16,513	01.08 %	566	0.65 %
Gooding	Gooding	1913	14,295	00.94 %	734	0.84 %
Idaho	Grangeville	1861/1864	15,448	01.01 %	8,502	9.71 %
Jefferson	Rigby	1913	23,860	01.57 %	1,106	1.26 %
Jerome	Jerome	1919	20,468	01.34 %	602	0.69 %
Kootenai	Coeur d'Alene	1864	137,475	09.02 %	1,316	1.50 %
Latah	Moscow	1886	35,906	02.36 %	1,077	1.23 %
Lemhi	Salmon	1869	7,808	00.51 %	4,570	5.22 %
Lewis	Nezperce	1911	3,594	00.24 %	480	0.55 %
Lincoln	Shoshone	1895	4,503	00.30 %	1,206	1.38 %
Madison	Rexburg	1914	37,456	02.46 %	473	0.54 %
Minidoka	Rupert	1913	18,645	01.22 %	763	0.87 %
Nez Perce	Lewiston	1861/1864	38,975	02.56 %	856	0.98 %
Oneida	Malad City	1864	4,130	00.27 %	1,202	1.37 %
Owyhee	Murphy	1863	10,877	00.71 %	7,697	8.79 %
Payette	Payette	1917	22,966	01.51 %	410	0.47 %

Power	American Falls	1913	7,683	00.50 %	1,443	1.65 %
Shoshone	Wallace	1861/1864	12,913	00.85 %	2,636	3.01 %
Teton	Driggs	1915	8,833	00.58 %	451	0.52 %
Twin Falls	Twin Falls	1907	74,284	04.87 %	1,928	2.20 %
Valley	Cascade	1917	8,862	00.58 %	3,734	4.27 %
Washington	Weiser	1879	10,206	00.67 %	1,474	1.68 %

Total Counties: 44. Total 2008 Population Est.: 1,523,816. Total Area: 87530 square miles (km^2).

Three counties were first designated as such by the Washington Territorial Legislature in 1861;[40] they were subsequently re-designated as Idaho counties in 1864. The 1861 Nez Perce county has since been broken up into Nez Perce, Lewis, Boundary, Benewah, Latah, Kootenai and Clearwater counties.

Idaho license plates begin with a county designation based on the first letter of the county's name. Where a letter is at the beginning of more than one name, a number accompanies precedingly in alphabetical order. This reflects an anomalous coincidental situation wherein 10 counties begin with B, seven with C and four with L, which is 21 of the 44 counties.

Politics

Presidential elections results

Year	Republican	Democratic
2008	**61.5%** *403,012*	36.1% *236,440*
2004	**68.38%** *409,235*	30.26% *181,098*
2000	**67.17%** *336,937*	27.64% *138,637*
1996	**52.18%** *256,595*	33.65% *165,443*
1992	**42.03%** *202,645*	28.42% *137,013*
1988	**62.08%** *253,881*	36.01% *147,272*
1984	**72.36%** *297,523*	26.39% *108,510*
1980	**66.46%** *290,699*	25.19% *110,192*
1976	**59.88%** *204,151*	37.12% *126,549*
1972	**64.24%** *199,384*	26.04% *80,826*
1968	**56.79%** *165,369*	30.66% *89,273*
1964	49.08% *143,557*	**50.92%** *148,920*
1960	**53.78%** *161,597*	46.22% *138,853*

After the Civil War, many Midwestern and Southern Democrats moved to the Idaho Territory. As a result, the early territorial legislatures were solidly Democrat-controlled. In contrast, most of the territorial governors were appointed by Republican presidents and were Republicans themselves. This led to sometimes-bitter clashes between the two parties, including a range war with the Democrats backing the sheepherders and the Republicans the cattlemen. That ended with the "Diamondfield" Jack Davis murder trial. In the 1880s, Republicans became more prominent in local politics.

Since statehood, the Republican Party has usually been the dominant party in Idaho, as there was a polar shift in social and political stance between the two parties, when the Democrats became more liberal and the Republicans more conservative. At one time, Idaho had two Democratic parties, one being the mainstream and the other called the Anti-Mormon Democrats, lasting into the early 20th century. In the 1890s and early 1900s, the Populist Party enjoyed prominence while the Democratic Party maintained a brief dominance in the 1930s during the Great Depression. Since World War II, most statewide elected officials have been Republicans. The last time the Democratic Party held a majority in either house of the state legislature was the House of Representatives in 1958 by one seat. However, Democrats did hold the governorship from 1971 to 1995, despite the state's Republican tilt.

Idaho Congressional delegations have also been generally Republican since statehood. Several Idaho Democrats have had electoral success in the House over the years, but the Senate delegation has been a Republican stronghold for decades. Several Idaho Republicans, including current Senator Mike Crapo, have won reelection to the Senate, but only Frank Church has won reelection as a Democrat. Church was the last Idaho Democrat to win a U.S. Senate race, in 1974. Walt Minnick's 2008 win in the First Congressional District was the state's first Democratic Congressional victory in 16 years.

In modern times, Idaho has been a reliably Republican state in presidential politics as well. It has not supported a Democrat for president since 1964. Even in that election, Lyndon Johnson defeated Barry Goldwater in the state by fewer than two percentage points, compared to a landslide nationally. In 2004, Republican George W. Bush carried Idaho by a margin of 38 percentage points and with 68.4% of the vote, winning in 43 of 44 counties. Only Blaine County, which contains the Sun Valley ski resort, supported John Kerry, who owns a home in the area. In 2008 Barack Obama's 36.1 percent[41] showing was the best for a Democratic presidential candidate in Idaho since 1976. However, Republican margins were narrower in 1992 and 1996.

In the 2006 elections, Republicans, led by gubernatorial candidate C. L. "Butch" Otter, won all the state's constitutional offices and retained both of the state's seats in the United States House of Representatives. However, Democrats picked up several seats in the Idaho Legislature, notably in the Boise area.[42]

Republicans lost one of the House seats in 2008 to Minnick, but Republican Jim Risch retained Larry Craig's Senate seat for the GOP by a comfortable margin.[43]

Further information: Political party strength in Idaho

Important cities and towns

Population > 100,000 (urbanized area)

- Boise (state capital)

Population > 50,000 (urbanized area)

- Idaho Falls – Location of the main offices of the Idaho National Laboratory
- Nampa – Home of Northwest Nazarene University
- Pocatello – Home of Idaho State University
- Meridian – Suburb of Boise, Fastest growing city in Idaho

Population > 30,000 (urbanized area)

- Caldwell – Home of the College of Idaho
- Coeur d'Alene – Home of North Idaho College, major tourist hub
- Lewiston – Home of Lewis-Clark State College, Seaport
- Twin Falls – Home of College of Southern Idaho, BASE jumping

Population > 10,000 (urbanized area)

- Ammon – Suburb of Idaho Falls
- Blackfoot – Home of the Idaho Potato Museum
- Burley
- Eagle – Suburb of Boise
- Garden City – Suburb of Boise

- Hayden
- Jerome
- Kuna – Suburb of Boise
- Moscow – Home of the University of Idaho
- Mountain Home – U.S. Air Force Base
- Post Falls
- Rexburg – Home of Brigham Young University-Idaho

Smaller Towns and Cities

- American Falls – First town to be entirely relocated
- Arco – First city to be lit by electricity generated from a nuclear power plant
- Bonners Ferry – Northernmost major town in Idaho
- Buhl – "Trout capital of the world"
- Bone- Population 2 But still has gas station
- Driggs – skiing (Grand Targhee)
- Eden
- Emmett
- Greenleaf
- Firth- Home of the Firth Cougars
- Fruitland
- Filer – Suburb of Twin Falls
- Hazelton
- Homedale – town's name was chosen from a hat
- Island Park – snowmobiling, world-class fishing
- Kimberly – Suburb of Twin Falls
- Kellogg – skiing (Silver Mountain Ski Resort)
- Malad City
- McCall – skiing (Brundage Ski Resort) and Recreation Payette Lake
- Middleton
- Montpelier bank robbed by the wild bunch
- Mullan – silver/lead/zinc mining
- New Meadows - at the 45th parallel north
- New Plymouth – first planned community in Idaho, third west of the Rocky Mountains
- Notus
- Oakley – famous pioneer town, home of many historic buildings
- Orofino – Clearwater County seat, site of Dworshak Dam, highest straight-axis dam in Western hemisphere
- Paris – Bear Lake County seat
- Parma
- Payette – Payette County seat
- Plummer- CDA tribal headquarters
- Preston- location of the 2004 film *Napoleon Dynamite* and the annual International Bed Races
- Rupert- Minidoka County seat
- Rigby – Television birthplace
- Salmon – Gateway to "River of No Return" (Salmon River)
- Sandpoint – skiing (Schweitzer Mountain Ski Resort) and Recreation Lake Pend Oreille
- Shelley – Home of the Russet SPUD
- Soda Springs – U.S.'s only captive geyser
- St. Anthony – sand dunes and several lava tubes

- St. Maries – Benewah County seat
- Stanley, Idaho – heart of the Sawtooth mountains
- Star, Idaho
- Sun Valley – Year-round resort with world-class skiing
- Wallace – Historic district and Shoshone County seat
- Weiser – Washington County seat, home of the National Oldtime Fiddlers' Contest
- Wilder
- Worley – casino

Coeur d'Alene

Pocatello

Post Falls

Idaho Falls

City of Rocks National Reserve

Craters of the Moon National Monument

- California National Historic Trail
- City of Rocks National Reserve
- Craters of the Moon National Monument
- Hagerman Fossil Beds National Monument
- Lewis and Clark National Historic Trail
- Minidoka National Historic Site

- Nez Perce National Historical Park
- Oregon National Historic Trail
- Yellowstone National Park

National Recreation Areas

- Hells Canyon National Recreation Area
- Sawtooth National Recreation Area

Hells Canyon National Recreation Area.

National Wildlife Refuges

- Bear Lake National Wildlife Refuge
- Camas National Wildlife Refuge
- Deer Flat National Wildlife Refuge
- Grays Lake National Wildlife Refuge
- Kootenai National Wildlife Refuge
- Minidoka National Wildlife Refuge

National Conservation Areas

- Snake River Birds of Prey National Conservation Area

State parks

- Bear Lake State Park
- Bruneau Dunes State Park
- Castle Rocks State Park
- City of Rocks National Reserve
- Coeur d'Alene Parkway State Park
- Dworshak State Park
- Eagle Island State Park
- Farragut State Park
- Harriman State Park
- Hells Gate State Park
- Henrys Lake State Park
- Heyburn State Park
- Lake Cascade State Park
- Lake Walcott State Park
- Land of the Yankee Fork State Park
- Lucky Peak State Park
- Massacre Rocks State Park
- McCroskey State Park
- Old Mission State Park
- Ponderosa State Park
- Priest Lake State Park

Bear lake viewed from Bear Lake State Park.

- Round Lake State Park
- Thousand Springs State Park
- Three Island Crossing State Park
- Trail of the Coeur d'Alenes
- Winchester Lake State Park

Education

Colleges and universities

The Idaho State Board of Education oversees three comprehensive universities. The University of Idaho in Moscow was the first university in the state (founded in 1889). A land-grant institution, the UI is the state's flagship university. Idaho State University in Pocatello opened in 1901 as the Academy of Idaho and received university status in 1963. Boise State University is the most recent school to attain university status in Idaho, and is primarily geared toward being a commuter school for part-time undergraduate students. The school opened in 1932 as Boise Junior College and became Boise State University in 1974. Lewis-Clark State College in Lewiston is the only public, non-university 4-year college in Idaho.

The Jacob Spori Building at Brigham Young University-Idaho in Rexburg.

Idaho has three regional community colleges: North Idaho College in Coeur d'Alene; College of Southern Idaho in Twin Falls; and College of Western Idaho in Nampa, which opened in 2009. A public technical college, Eastern Idaho Technical College, operates in Idaho Falls.

Idaho State University in Pocatello.

Private institutions in Idaho are Boise Bible College, affiliated with congregations of the Christian churches and churches of Christ; Brigham Young University-Idaho in Rexburg, which is affiliated with the Church of Jesus Christ of Latter-day Saints; The College of Idaho in Caldwell, which still maintains a loose affiliation with the Presbyterian Church; Northwest Nazarene University in Nampa; and New Saint Andrews College in Moscow, of reformed Christian theological background.

University of Idaho in Moscow.

Boise State University in Boise.

- Boise Bible College
- Boise State University
- Brigham Young University-Idaho
- College of Idaho
- College of Southern Idaho
- College of Western Idaho
- Eastern Idaho Technical College
- Idaho State University
- Lewis-Clark State College
- New Saint Andrews College
- North Idaho College
- Northwest Nazarene University
- University of Idaho

Sports

Central Idaho is home to one of North America's oldest ski resorts, Sun Valley, where the world's first chairlift was installed. Sites include Hell's Canyon, the Salmon River, and the embarkation point of Riggins, ID.

Club	Sport	League
Boise Hawks	Baseball	Minor League Baseball
Boise State Broncos	NCAA	Div I FBS – MW
Idaho Vandals	NCAA	Div I FBS – WAC
Idaho State Bengals	NCAA	Div I FCS – Big Sky
Idaho Falls Chukars	Baseball	Minor League Baseball
Idaho Stampede	Basketball	NBA Development League
Idaho Steelheads	Ice hockey	East Coast Hockey League

Boise is the host to the largest 5 km run for women, the St. Luke's Women's Fitness Celebration.

Official state emblems

- State Bird: Mountain Bluebird[44]
- State Dance: Square Dance[44]
- State Fish: Cutthroat Trout[44]
- State Flower: Syringa (*Philadelphus lewisii*)[44]
- State Fossil: Hagerman Horse (*Equus simplicidens*)[44]
- State Fruit: Huckleberry[44]
- State Gem: Idaho star garnet[44]
- State Horse: Appaloosa[44]
- State Motto: *Esto perpetua* ("Let it be perpetual")[44] [45]

- State Insect: Monarch butterfly[44]
- State Raptor: Peregrine falcon[44]
- State Song: Here We Have Idaho[44]
- State Tree: Western White Pine[44]
- State Vegetable: Potato[44]

See also

- List of National Register of Historic Places in Idaho
- List of people from Idaho
- List of places in Idaho
- Rocky Mountains
- US state

References

[1] "Beauty Reset" (http://www.ngs.noaa.gov/cgi-bin/ds_mark.prl?PidBox=PZ0770). *NGS data sheet*. U.S. National Geodetic Survey. . Retrieved October 20, 2011.
[2] "Elevations and Distances in the United States" (http://egsc.usgs.gov/isb/pubs/booklets/elvadist/elvadist.html). United States Geological Survey. 2001. . Retrieved October 21, 2011.
[3] Elevation adjusted to North American Vertical Datum of 1988.
[4] http://www.idaho.gov/
[5] Just, Rick. "Star Garnet." *Idaho Snapshots*. Meridian, Idaho: Radio Idaho, 1990. 9.
[6] "Idaho-its meaning, origin and application : Rees, John E., 1868–1928 : Free Download & Streaming : Internet Archive" (http://www.archive.org/details/idahoitsmeaningo00reesrich). Archive.org. . Retrieved 2010-07-30.
[7] "Idaho – MSN Encarta" (http://encarta.msn.com/encyclopedia_761565515/Idaho.html). *Idaho – MSN Encarta*. .
[8] "Did Idaho Get Its Name As A Result Of A Hoax?" (http://www.museumofhoaxes.com/hoax/weblog/permalink/did_idaho_get_its_name_as_a_result_of_a_hoax/). Museumofhoaxes.com. 2006-04-25. . Retrieved 2010-07-30.
[9] (http://www.etymonline.com/index.php?term=Idaho) *etymonline.com*
[10] Barber & Martin (1956). *Idaho in the Pacific Northwest*. Caxton Printers Ltd. Library of Congress 55-5192.
[11] "Western States Data Public Land Acreage" (http://www.wildlandfire.com/docs/2007/western-states-data-public-land.htm). Wildlandfire.com. 2007-11-13. . Retrieved 2010-07-30.
[12] "USDA Forest Service – Comment Form" (http://www.fs.fed.us/land/staff/lar/2007/Table_4.htm). Fs.fed.us. 2005-04-01. . Retrieved 2010-07-30.
[13] "Port of Lewiston" (http://www.u-s-history.com/pages/h2812.html). u-s-history.com. . Retrieved 2010-07-30.
[14] "Sawtooth Range" (http://www.idahoaclimbingguide.com/id27.htm). Idahoaclimbingguide.com. . Retrieved 2010-07-30.
[15] http://www.law.cornell.edu/uscode/15/usc_sup_01_15_10_6.html
[16] "264, Part of Idaho in fourth zone" (http://uscode.house.gov/download/pls/15C6.txt). .
[17] "Climate Of Idaho" (http://www.wrcc.dri.edu/narratives/IDAHO.htm). Wrcc.dri.edu. 1954-02-20. . Retrieved 2010-07-30.
[18] http://www.ustravelweather.com/weather-idaho/
[19] "David Thompson's Trading Post, Idaho Forts, American Forts Network" (http://www.northamericanforts.com/West/id.html#thompson). .
[20] Meinig, D.W. (1995) [1968]. *The Great Columbia Plain* (Weyerhaeuser Environmental Classic ed.). University of Washington Press. pp. 36, 55. ISBN 0-295-97485-0.
[21] "Fur Trade Posts In Idaho, Idaho State Historical Society." (http://www.idahohistory.net/Reference Series/0062.pdf). .
[22] "Donald MacKenzie's Post, Idaho Forts, American Forts Network" (http://www.northamericanforts.com/West/id.html#clear). .
[23] Bennett, Eldon T.. "An Early History of Franklin" (http://www.franklinidaho.org/History2.htm). Franklin, Idaho. . Retrieved 2008-05-19.
[24] "Elias Davidson Pierce and the Founding of Pierce" (http://www.idahohistory.net/Reference Series/0008.pdf) (PDF). Idaho State Historical Society. August 1966. . Retrieved 2008-05-19.
[25] ""Mormon" Entry for The Encyclopedia of the Supreme Court of the United States, David S.Tanenhaus" (http://nboman.people.wm.edu/MormonsEUSC.pdf) (PDF). . Retrieved 2010-07-30.

[26] "The Power of Idaho" (http://ieda.biz/white.html). Idaho Economic Development Association. 2004. . Retrieved 2007-10-07.
[27] http://2010.census.gov/2010census/data/apportionment-pop-text.php
[28] "Idaho QuickFacts from the U.S. Census Bureau" (http://quickfacts.census.gov/qfd/states/16000.html). Quickfacts.census.gov. . Retrieved 2010-07-30.
[29] "Population and Population Centers by State" (http://www.census.gov/geo/cenpop/statecenters.txt). United States Census Bureau. 2000. . Retrieved 2008-12-04.
[30] "RLS report 2-22.indd" (http://religions.pewforum.org/pdf/report-religious-landscape-study-full.pdf) (PDF). . Retrieved 2010-07-30.
[31] "State Membership Report – Idaho" (http://www.thearda.com/mapsReports/reports/state/16_2000.assp). Association of Religion Data Archives. . Retrieved 2009-12-13.
[32] "Zuivelzicht" April 25, 2007
[33] "Today in History: March 4" (http://memory.loc.gov/ammem/today/mar04.html). Memory.loc.gov. . Retrieved 2010-07-30.
[34] "Facts At a Glance" (http://web.archive.org/web/20070613232433/http://idaholottery.com/facts.asp). Idaho Lottery. 2007. Archived from the original (http://idaholottery.com/facts.asp) on June 13, 2007. . Retrieved 2007-04-29.
[35] "2007 Idaho Energy Plan" (http://www.energy.idaho.gov/informationresources/d/energy_plan_2007.pdf). Idaho Legislative Council Interim Committee on Energy, Environment and Technology. 2007. . Retrieved 2010-11-30.
[36] "Idaho Energy Profile" (http://tonto.eia.doe.gov/state/state_energy_profiles.cfm?sid=ID). Energy Information Administration. 2009. . Retrieved 2007-06-02.
[37] "Idaho Strategic Energy Alliance Frequently Asked Questions" (http://www.energy.idaho.gov/idahostrategicenergyalliance/d/isea_faq.pdf). Idaho Strategic Energy Alliance. 2009. . Retrieved 2007-06-02.
[38] "Idaho Energy Complex" (http://idahoenergycomplex.org/faq.htm). Idaho Energy Complex. 2009. . Retrieved 2007-06-02.
[39] "Idaho District Court Websites" (http://www.isc.idaho.gov/district.htm). Isc.idaho.gov. . Retrieved 2008-12-17.
[40] "IdahoHistory.net" (http://www.idahohistory.net/pierce.html). IdahoHistory.net. 2010-07-07. . Retrieved 2010-07-30.
[41] Idaho Secretary of State Election Division, "November 4, 2008 General Election Results" (http://www.idsos.state.id.us/elect/RESULTS/2008/general/tot_stwd.htm)
[42] (http://www.ncsl.org/ncsldb/elect98/profile.cfm?yearsel=2006&statesel=ID)
[43] 2008 statewide totals (http://www.sos.idaho.gov/ELECT/results/ENR/statewide_total.html)
[44] "Governor C. L. "Butch" Otter – The State of Idaho – FYI Idaho State Symbols" (http://gov.idaho.gov/fyi/symbols/symbols_index.html). Gov.idaho.gov. . Retrieved 2010-07-30.
[45] "Idaho history homepage" (http://idahohistory.net/Reference Series/0134.pdf). Idahohistory.net. 2010-07-07. . Retrieved 2010-07-30.

External links

- State of Idaho government website (http://www.idaho.gov/)
- Idaho (http://www.dmoz.org/Regional/North_America/United_States/Idaho/) at the Open Directory Project
- Energy Profile for Idaho (http://tonto.eia.doe.gov/state/state_energy_profiles.cfm?sid=ID)
- Idaho Department of Parks and Recreation (http://www.parksandrecreation.idaho.gov/)
- Idaho Newspapers (http://www.usnewspapers.org/state/idaho)
- Idaho State Databases (http://wikis.ala.org/godort/index.php/Idaho) – Annotated list of searchable databases produced by Idaho state agencies and compiled by the Government Documents Roundtable of the American Library Association.
- Idaho State Facts (http://www.ers.usda.gov/StateFacts/ID.htm)
- Log Cabins in America: The Finnish Experience, a National Park Service Teaching with Historic Places (TwHP) lesson plan (http://www.nps.gov/history/NR/twhp/wwwlps/lessons/4logcabins/4logcabins.htm)
- The History of Idaho (http://gov.idaho.gov/fyi/history/history_index.html)
- U.S. Census Bureau (http://quickfacts.census.gov/qfd/states/16000.html)
- USGS real-time, geographic, and other scientific resources of Idaho (http://www.usgs.gov/state/state.asp?State=ID)
- Visit Idaho site (http://www.visitidaho.org/) – The official state tourism website

Related information

gag:Idaho mrj:Айдахо

Bitterroot Range

This article is about the entire Bitterroot Range. For the Bitterroot Mountains, see its section below or the Bitterroot Mountains article.

Bitterroot Range	
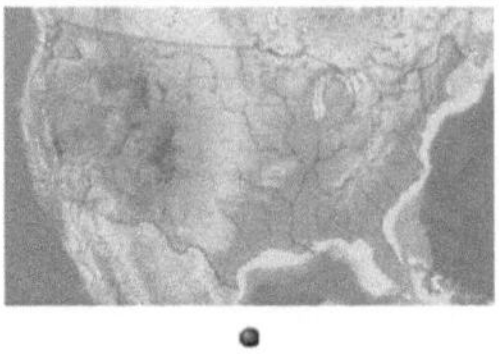	
Location	Idaho Montana USA
Nearest city	Missoula, MT
Coordinates	46°00′00″N 114°30′00″W

The **Bitterroot Range** (a subrange of the Rocky Mountains) runs along the border of Montana and Idaho in the northwestern United States. The range spans an area of 62,736 square kilometers (24,223 sq mi) and is named after the bitterroot (*Lewisia rediviva*), a small pink flower that is the state flower of Montana.[1]

History

In 1805, the Corps of Discovery, led by Meriwether Lewis and William Clark and aided by the Shoshone Native American tribe, crossed the Bitterroot Range several times. Lewis first crossed the mountains at Lemhi Pass on August 12, then returned across the pass to meet Clark. The entire expedition then crossed the pass to the Salmon River valley, and the next month entered the Bitterroot Valley from the south via either Lost Trail Pass or Chief Joseph Pass. It then crossed Lolo Pass to the west.

Ranges

According to the U.S. Board on Geographic Names, the Bitterroot Range runs from Pend Oreille Lake (near Sandpoint, Idaho) to Monida Pass,.[2] It is sometimes considered to extend east of the Monida Pass to include the Centennial Mountains. The range comprises the following subranges (from north to south):

Part of the Bitterroot Range in Montana, looking north from El Capitan peak

Coeur d'Alène Mountains

The **Coeur d'Alène Mountains** are the northwestern-most portion of the Bitterroot Range and encompass an area of 2,590 square miles (6,708 km²). The mountain range's two highest peaks are the 7,352 foot (2,241 m) Cherry Peak and the 6,837 foot (2,084 m) Patricks Knob.[3]

Saint Joe Mountains

The **Saint Joe Mountains**, the smallest named portion of the Bitterroot Range, encompass an area of 698 square miles (1,808 km²). They lie between the St. Joe River on the south, the Coeur d'Alène River on the north, the Slate Creek saddle on the east and the plateau of the Moscow, Idaho/Pullman, Washington area on the Idaho/Washington border.[4] [5]

Bitterroot Mountains

The **Bitterroot Mountains**, comprising the Northern and Central Bitterroot Ranges, are the largest portion of the Bitterroot Range and encompass an area of 4,862 square miles (12,593 km²). The mountains are bordered on the north by Lolo Creek, on the south by the Salmon River, on the east by the Bitterroot River and Valley, and on the west by the Selway and Lochsa Rivers.[6] Its highest summit is Trapper Peak, at 10,157 feet (3,096 m).[7] [8]

Beaverhead Mountains

The **Beaverhead Mountains** encompass an area of 4,532 square miles (11,738 km²). They lie to the east of the Bitterroot Mountains and lie to the west of the Big Hole Basin and the Pioneer Mountains. Passes in the mountains include Lemhi Pass, Bannock Pass, Big Hole Pass, Big Hole Pass II, Junction Pass and Monida Pass. The Beaverheads are further subdivided into the West Big Hole Mountains, the Big Hole Divide, the Tendoy Mountains, the Italian Peaks, and the Garfield Peaks.[9]

Centennial Mountains

The **Centennial Mountains** encompass an area of 2,064 square miles (5,346 km²).[10]

The Centennials are home to Brower's Spring, discovered in 1888 by Jacob V. Brower, which is believed to be the furthest point on the Missouri River. Brower published his finding in 1896 in "The Missouri: It's Utmost Source."

The site of Brower's Spring at around 8,800 feet (2,680 m) feet in the Centennials. The site now commemorated by a rock pile at the source of Hellroaring Creek which flows into Red Rock River and then into Clark canyon reservoir where it joins the Beaverhead River and then the Big Hole River before ultimately hooking up with the Jefferson River.[11]

References

[1] "Bitterroot Range" (http://www.bivouac.com/ArxPg.asp?ArxId=1742). *Canadian Mountain Encyclopedia*. Bivouac.com. Retrieved 4 March 2007.

[2] "Bitterroot Range" (http://geonames.usgs.gov/pls/gnispublic/f?p=gnispq:3:::NO::P3_FID:395685). Geographic Names Information System, U.S. Geological Survey. . Retrieved 2010-07-10.

[3] "Couer d" (http://www.peakbagger.com/range.aspx?rid=14310). Peakbagger.com. . Retrieved 2011-08-08'Alène Mountains. Retrieved 4 March 2007.

[4] "Saint Joe Mountains" (http://www.peakbagger.com/range.aspx?rid=14311). Peakbagger.com. . Retrieved 2011-08-08.. Retrieved 4 March 2007.

[5] St. "Joe Mountains" (http://www.bivouac.com/ArxPg.asp?ArxId=1741). *Canadian Mountain Encyclopedia*. Bivouac.com. Retrieved 4 March 2007.

[6] "Bitterroot Mountains" (http://www.bivouac.com/ArxPg.asp?ArxId=1739). *Canadian Mountain Encyclopedia* Bivouac.com. Retrieved 4 March 2007.

[7] "Northern Bitterroot Range" (http://www.peakbagger.com/range.aspx?rid=14312). Peakbagger.com. . Retrieved 2011-08-08. Retrieved 4 March 2007.

[8] "Central Bitterroot Range" (http://www.peakbagger.com/range.aspx?rid=14313). Peakbagger.com. . Retrieved 2011-08-08. Retrieved 4 March 2007.

[9] "Beaverhead Mountains" (http://www.peakbagger.com/range.aspx?rid=14314). Peakbagger.com. . Retrieved 2011-08-08. Retrieved 4 March 2007.

[10] "Centennial Mountains" (http://www.peakbagger.com/range.aspx?rid=14315). Peakbagger.com. . Retrieved 2011-08-08. Retrieved 4 March 2007.

[11] The True Utmost Reaches of the Missouri - Montana Outdoors - July-August 2005 (http://fwp.mt.gov/mtoutdoors/HTML/articles/2005/MissouriSource.htm)

External links

- Perspective aerial image of the Bitterroot Range and the Bitterroot Valley (http://130.166.124.2/montana_panorama_atlas/page27/files/page27-1003-full.html)

Bitterroot National Forest

Bitterroot National Forest	
IUCN Category VI (Managed Resource Protected Area)	

Location	Ravalli / Missoula counties, Montana; Idaho County, Idaho, USA
Nearest city	Missoula, MT
Coordinates	46°13′58″N 113°57′49″W
Area	1,587,070 acres (6,422.64 km²)

Wildfire in the forest as photographed on August 6, 2000.

Bitterroot National Forest comprises 1.587 million acres (6,423 km²) in west-central Montana and eastern Idaho, of the United States. It is located primarily in Ravalli County, Montana (70.26% of the forest), but also has acreage in Idaho County, Idaho (29.24%), and Missoula County, Montana (0.49%).[1]

Founded in 1898, the forest is located in the Bitterroot and the Sapphire mountain ranges with elevations ranging from 2,200 feet (650 m) along the Salmon River in Idaho to 10,157 foot (3,100 m) Trapper Peak. Roughly half the forest (743,000 acres, 3,000 km²) make up part or all of three distinct Wilderness areas. These areas include the Anaconda-Pintler, Selway-Bitterroot and Frank Church River of No Return Wildernesses. The distinction is that in wilderness areas, no roads, logging, mining or other construction is permitted and all access must be done either on foot or horseback; even bicycles are not permitted. Hunting however is allowed forest-wide including wilderness areas.

History

Trapper Peak in Bitterroot National Forest

The Lewis and Clark Expedition passed through parts of what are now forest lands in 1805. After the discovery of gold in Idaho and then Montana in the 1860s, numerous mining towns were built, some of which today are ghost towns. The Nez Perce National Historic Trail passes through a portion of the forest, following the route of the retreating Nez Perce on their historic path that led from Idaho to north central Montana in 1877. Heavy logging and other resource depletion beginning in the 1880s led conservationists to push for the preservation the forest.[2]

The **Bitter Root Forest Reserve** was established by the General Land Office on March 1, 1898 with 4147200 acres (16783 km^2). It was transferred to the U.S. Forest Service in 1906. On July 1, 1908 the name was changed to Bitterroot National Forest, with lands added from Big Hole National Forest and Hell Gate National Forest. Other lands were transferred from Biterroot to Beaverhead, Clearwater, Nez Perce and Salmon National Forests. On October 29, 1934 part of Selway National Forest was added.[3]

Composition

The forest is a combination of both grasslands and forested zones. Grazing rights are leased to private landowners in the lower altitudes where grasses and shrublands are dominant. Higher up, Douglas fir, larch, and lodgepole pine slowly give way to Engelmann Spruce and whitebark pine as the altitude increases. Above the treeline at 8,000 feet (2,400 m) the trees abruptly end and alpine flowers and grasses are found. A small grizzly bear population is located in the wilderness zones of the forest with black bear, mountain goat, bighorn sheep, elk and moose found forestwide. An active effort to reintroduce the grizzly bear to the region concluded in 2000 with a plan to release 25 bears into the wilderness zones over a five year period beginning in 2003.[4]

There are 1,600 mi (2,500 km) of trails and 18 improved campgrounds within the forest. Outstanding fishing is found in the dozens of rivers and streams and lakes. The forest headquarters is located in Hamilton, Montana. There are local ranger district offices in Darby, Stevensville, and Sula.[5] The largest nearby city is Missoula, Montana. The scenic Blodgett Canyon is but one of many steep canyons located in the forest. U.S. Highway 93 passes through portions of the forest.

Bitterroot controversy

Swanson (2011) examines the critical role of Guy M. Brandborg of the U.S. Forest Service, who was supervisor of the Bitterroot National Forest from 1935 to 1955. By insisting on selective cutting, his tried to protect the watersheds and wildlife habitats that are harmed by clear-cutting. After he retired in 1955 Brandborg denounced the Forest Service for deviating from his model of sustainable forestry. He launched a public attack, known as the "Bitterroot controversy." Brandborg lobbied to secure passage of the National Forest Management Act of 1976, that codified his ideals.

Further reading

- Swanson, Frederick H. *The Bitterroot and Mr. Brandborg: Clearcutting and the Struggle for Sustainable Forestry in the Northern Rockies* (University of Utah Press, 2011). ISBN 978-1-60781-101-5

See also

- 2000–2001 fires in the Western United States
- Bitterroot Mountains
- List of Forests in Montana

References

[1] Table 6 - NFS Acreage by State, Congressional District, and County, 30 September 2008 (http://www.fs.fed.us/land/staff/lar/2008/TABLE_6.htm)

[2] Wilderness.net. "Selway-Bitterroot Wilderness" (http://www.wilderness.net/index.cfm?fuse=NWPS&sec=wildView&wname=Selway-Bitterroot). *The National Wilderness Preservation System.* . Retrieved 2006-07-08.

[3] Davis, Richard C. (September 29, 2005). *National Forests of the United States* (http://www.foresthistory.org/ASPNET/Places/National Forests of the U.S.pdf)*PDF (341 KB)*. The Forest History Society.

[4] U.S. Fish and Wildlife Service. "Record of Decision and Final Rule for Grizzly Bear Recovery in the Bitterroot" (http://www.r6.fws.gov/bitterroot/). . Retrieved 2006-07-08.

[5] USFS Ranger Districts by State (http://www.ufwda.org/pdfs/USDAForestServiceRangerDistricts.pdf)

External links

- Bitterroot National Forest (http://www.fs.fed.us/r1/bitterroot/) - U.S. Forest Service
- USGS Gird Point (MT) Topo Map (http://www.topoquest.com/map.asp?lat=46.23293&lon=-113.96352&s=500&size=l&u=4&datum=nad27&layer=DRG) - TopoQuest.com
- Bitterroot National Forest Recreation (http://explorethebitterroot.com/bitterroot-national-forest)

Clearwater National Forest

Clearwater National Forest	
IUCN Category VI (Managed Resource Protected Area)	

Location	Idaho, USA
Nearest city	Orofino, ID
Coordinates	46°33′4″N 115°9′37″W
Area	2,612,000 acres (10,570 km²)

Clearwater National Forest is located in North Central Idaho in the northwestern United States. The forest is bounded on the east by the state of Montana, on the north by the Idaho Panhandle National Forest, and on the south and west by the Nez Perce National Forest and Palouse Prairie.

Lochsa River Canyon
Courtesy of Idaho Travel Council [1]

The forest is 1.8 million acres (7,300 km²) and contains mountains and river drainage areas. The Lochsa and Clearwater Rivers flow through the forest region. Portions of the Selway-Bitterroot Wilderness are also located in the forest. In descending order of land area the forest is located in Idaho, Clearwater, and Shoshone counties. Forest Service headquarters are located in Orofino, Idaho. There are local ranger district offices in Kamiah, Kooskia, Lolo (Montana), Orofino, and Potlatch (all in Idaho, except as indicated).[2]

The Clearwater National Forest is home to the well-known Jerry Johnson Hot Springs and the smaller Weir Creek Hot Springs. Both springs are accessible via trails from U.S. Highway 12 and are popular tourist attractions, drawing visitors from Idaho and western Montana.

Lewis and Clark travelled through the region in their journey westward in 1805-1806. Upon reaching Lolo Pass at the eastern edge of the forest on September 15, 1805, William Clark was quoted as saying,

> "From this mountain I could observe high rugged mountains in every direction as far as I could see."

The Clearwater National Forest was established on July 1, 1908 with 2687860 acres (10877.4 km^2) from parts of Coeur d'Alene National Forest and Bitterroot National Forest. On October 29, 1934 part of Selway National Forest was added.[3]

References

[1] http://www.visitid.org/
[2] USFS Ranger Districts by State (http://www.ufwda.org/pdfs/USDAForestServiceRangerDistricts.pdf)
[3] Davis, Richard C. (September 29, 2005), *National Forests of the United States* (http://www.foresthistory.org/ASPNET/Places/National Forests of the U.S.pdf), Forest History Society,

External links

- Official site (http://www.fs.fed.us/r1/clearwater/)

Lolo National Forest

Lolo National Forest	
IUCN Category VI (Managed Resource Protected Area)	

Location	Montana, USA
Nearest city	Missoula, MT
Coordinates	47°09′0″N 114°27′0″W
Area	2 million acres (8,000 km^2)

Lolo National Forest is located in western Montana, United States with the western boundary being the state of Idaho. The forest spans 2 million acres (8,000 km^2) and includes four wilderness areas; the Scapegoat and the Bob Marshall Wilderness are partially within the forest while the Welcome Creek and Mission Mountains Wildernesses are solely in Lolo National Forest. The forest was created in 1906 from 4 different previous forests which were combined for administrative purposes. Lolo National Forest is west of the Continental divide and has a biodiversity influenced by both continental and maritime weather creating a transitional forest that has a high number of different plant and tree species.

Western red cedar, larch and whitebark pine share the forest with a variety of spruce and fir tree species. Western red cedars grow larger in Lolo National Forest than any other tree species does anywhere in Montana, attaining over 8 feet (250 cm) in diameter and almost 200 feet (60 m) in height. In total, 1,500 plant species exist in the forest as well as 60 species of mammals, 20 varieties of fish and 300 species of birds. Large mammals found in Lolo National Forest include the grizzly, black bear, mountain goat, bighorn sheep, elk, moose and mule deer. Both the bald and golden eagle reside in the forest along with trumpeter swan, herons and 30 varieties of ducks.

Rainy Lake in Lolo National Forest

The forest has a dozen improved campgrounds and 700 miles (1,100 km) of hiking trails. There are over 100 named lakes and five rivers including the Flathead River. The city of Missoula, Montana is the location for the forest headquarters and is centrally positioned within sight of the forestlands. There are local ranger district offices in Huson, Missoula, Plains, Seeley Lake, and Superior.[1] The tallest point in the Lolo is Scapegoat Mountain at 9186 feet (2800 m).

In descending order of land area the forest is located in parts of Mineral, Missoula, Sanders, Granite, Powell, Lewis and Clark, Flathead, and Ravalli counties.

Lolo National Forest was established on September 20, 1906 with 1211680 acres (4903.5 km^2). On December 16, 1931 part of Missoula National Forest was added, followed by portions of Selway National Forest in 1934 and part of Cabinet National Forest in 1954.[2]

Special Places

- Ninemile Wildlands Training Center - Located near Frenchtown, Ninemile Remount Depot listed on the National Register of Historic Places
- Lolo National Historic Trail - Historic network of trails traveled by the Nez Perce and Lewis and Clark.
- Savenac Historic Tree Nursery - Located near Haugan, former USDA Forest Service nursery designated a National Historic Site

See also

- List of Forests in Montana

References

[1] USFS Ranger Districts by State (http://www.ufwda.org/pdfs/USDAForestServiceRangerDistricts.pdf)
[2] Davis, Richard C. (September 29, 2005). *National Forests of the United States* (http://www.foresthistory.org/ASPNET/Places/National Forests of the U.S.pdf)*PDF (341 KB)*. The Forest History Society.

External links

- "Lolo National Forest" (http://www.fs.fed.us/r1/lolo/about/). U.S. Forest Service. Retrieved 2006-07-08.
- "Lolo National Forest: Special Places" (http://fs.usda.gov/wps/portal/fsinternet/!ut/p/c4/04_SB8K8xLLM9MSSzPy8xBz9CP0os3gDfxMDT8MwRydLA1cj72DTgBAjAwjQL8h2VAQAU_HvEQ!!/?ss=110116&navtype=BROWSEBYSUBJECT&cid=FSE_003693&navid=100000000000000&pnavid=null&position=BROWSEBYSUBJECT&ttype=attmain&pname=Lolo National Forest- Special Places). Lolo National Forest.
- "Lolo National Historic Trail" (http://fs.usda.gov/wps/portal/fsinternet/!ut/p/c4/04_SB8K8xLLM9MSSzPy8xBz9CP0os3gDfxMDT8MwRydLA1cj72DTUE8TAwjQL8h2VAQAMtzFUw!!/?navtype=BROWSEBYSUBJECT&cid=fsm9_021410&navid=150140000000000&pnavid=150000000000000&ss=110116&position=Not Yet Determined.Html&ttype=detailfull&pname=Lolo National Forest - History). Lolo National Forest.

Frank Church—River of No Return Wilderness

Frank Church-River of No Return Wilderness	
IUCN Category Ib (Wilderness Area)	

Location	Idaho / Valley / Lemhi counties, Idaho, USA
Nearest city	Yellow Pine, ID
Coordinates	45°3′55″N 114°57′16″W
Area	2,366,757 acres (9,577 km²)

The **Frank Church—River of No Return Wilderness Area** is a protected wilderness area in Idaho. It was created in 1980 by the United States Congress as the *River of No Return Wilderness Area*, and renamed in 1984 as the *Frank Church-River of No Return Wilderness Area* in honor of U.S. Senator Frank Church.

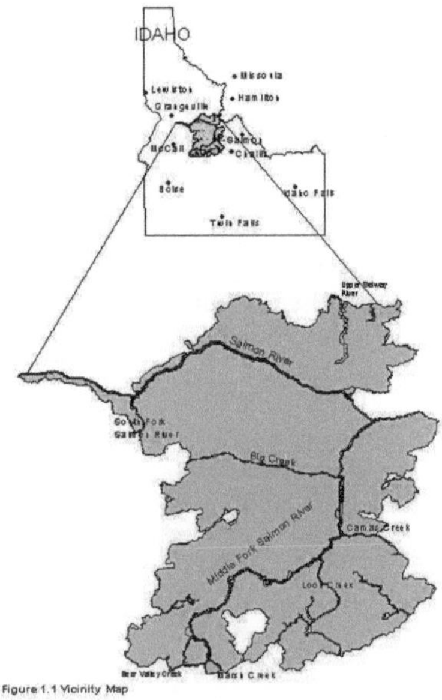

Map of Idaho showing the Frank Church-River of No Return Wilderness

At 2.3 million acres (9,300 km²), it is the second largest contiguous area of protected wilderness in the continental United States, after Death Valley.[1] Together with the adjacent Gospel Hump Wilderness and surrounding unprotected roadless Forest Service land, it is the core of a 3.3 million acre (13,000 km²) roadless area. It is separated from the Selway-Bitterroot Wilderness, to the north, by a single dirt road (the Magruder Corridor). The wilderness contains parts of several mountain ranges, including the Salmon River Mountains, the Clearwater Mountains, and the Bighorn Crags. The ranges are split by steep canyons of the Middle and Main forks of the Salmon River. The Salmon River is a popular destination for whitewater rafting, and is colloquially known as the "River of No Return" for its swift current which makes upstream travel difficult. Most of the area is covered by coniferous forests, with dry, open land along the rivers at lower elevations.

While designation as a Wilderness Area in the United States generally requires the prohibition of any motorized machinery, the use of jetboats (On the Main Fork of the Salmon River) and several airstrips are permitted in this wilderness as grandfathered existing uses before the wilderness was designated.

History

In 1931, 1,090,000 acres (4,400 km^2) in Central Idaho were declared by the U.S. Forest Service as The Idaho Primitive Area. In 1963, the Selway-Bitterroot Wilderness was split into three parts: The Selway-Bitterroot Wilderness, the Salmon River Breaks Primitive area, and the Magruder Corridor-the land between the two areas.

Frank Church was the Senate floor sponsor for The Wilderness Act of 1964, which protected 9 million acres (36,000 km^2) of United States land as part of the National Wilderness Preservation System. In 1968, he introduced the Wild and Scenic Rivers Act, which included the Middle Fork of the Salmon River, so that rivers "shall be preserved in free-flowing condition, and that they and their immediate environments shall be protected for the benefit and enjoyment of present and future generations."

Church's environmental legislation culminated in 1980 with the passage of the Central Idaho Wilderness Act. The act created the River of No Return Wilderness by combining the Idaho Primitive Area, the Salmon River Breaks Primitive Area, and a portion of the Magruder Corridor. The Act also added 125 miles (201 km) of the Salmon River to the Wild and Scenic Rivers System. In March 1984, weeks before his death, Congress honored Senator Church by renaming the area, The Frank Church—River of No Return Wilderness.

Wildlife

Because of its size the wilderness area provides a secluded habitat for a wide variety of mammal species, including some rare, vulnerable species. The wilderness is inhabited by a large population of mountain lions and grey wolves that visit the area. Populations of black bears, as well as: lynx, coyote, and red fox are scattered throughout the area. Other observable ruminant wildlife within the wilderness include: bighorn sheep, mountain goats, elk, moose, mule deer, and white tail deer. While this area has been deemed as one of the few remaining areas in the continental U.S. with suitable habitat for grizzly bears, no established populations are known to exist. The wilderness also offers some of the most critical habitat for wolverines in the lower 48 states.

See also

- List of U.S. Wilderness Areas
- Wilderness Act

Notes

[1] US Forest Service - Caring for the land and serving people. (http://www.fs.fed.us/r4/sc/fcronr/fcronrindex.shtml)

Bibliography

- Dant, Sara. "Making Wilderness Work: Frank Church and the American Wilderness Movement." Pacific Historical Review 77 (May 2008): 237-272.

External links

- US Forest Service Page on the Frank Church-RNR Area (http://www.fs.fed.us/r4/sc/recreation/fcronr/index.shtml)
- Wilderness.net wilderness information (http://www.wilderness.net/index.cfm?fuse=NWPS&sec=wildView&wname=Frank Church-River of No Return)
- TopoQuest map of the wilderness (http://www.topoquest.com/map.asp?z=11&n=5022428&e=701319&s=100&size=l&u=7&datum=nad83&layer=DRG25)
- Wolverines: Ghosts In Our Woods (http://www.svguide.com/w06/w06_wolverines.htm) from Sun Valley Guide

Nez Perce National Forest

Nez Perce National Forest	
IUCN Category VI (Managed Resource Protected Area)	

Location	Idaho County, Idaho, USA
Nearest city	Grangeville, Idaho
Coordinates	45°27′00″N 115°55′03″W
Area	2,224,091 acres (9,000.58 km^2)

The **Nez Perce National Forest** is located in Idaho County in central western Idaho in the northwestern United States. The forest is bounded on the east by the state of Montana, on the north by the Palouse region, and on the south and west by the Clearwater National Forest. It has a total area of 2,224,091 acres (3,475.14 sq mi or 9,000.58 km^2).[1] Forest headquarters are located in Grangeville, Idaho. There are local ranger district offices in Elk City, Grangeville, Kooskia, and White Bird.[2] It is the largest of the six National Forests that are contained entirely within a single county.

Nez Perce was established on July 1, 1908 by the U.S. Forest Service with 1946340 acres (7876.6 km^2) from parts of Bitter Root National Forest and Weiser National Forest. On October 29, 1934 part of Selway National Forest was added.[3]

Crooked Creek in Gospel Hump Wilderness

See also

- Nez Perce tribe

References

[1] Table 6 - NFS Acreage by State, Congressional District and County (http://www.fs.fed.us/land/staff/lar/2007/TABLE_6.htm) - United States Forest Service - September 30, 2007

[2] USFS Ranger Districts by State (http://www.ufwda.org/pdfs/USDAForestServiceRangerDistricts.pdf)

[3] Davis, Richard C. (September 29, 2005), *National Forests of the United States* (http://www.foresthistory.org/ASPNET/Places/National Forests of the U.S.pdf), The Forest History Society,

External links

- Official site (http://www.fs.fed.us/r1/nezperce/)

Lochsa River

Lochsa River	
 Headwaters near Powell	
Name origin: Nez Perce word for "rough water"	
Country	United States
State	Idaho
County	Idaho
Source	Confluence of Crooked Fork and Colt Killed Creek, Bitterroot Mountains
- location	Near Powell Ranger Station, Idaho County, Idaho
- elevation	3441 ft (1049 m) [1]
- coordinates	46°30′30″N 114°40′54″W [2]
Mouth	Clearwater River
- location	Lowell, Idaho County, Idaho
- elevation	1453 ft (443 m) [2]
- coordinates	46°08′25″N 115°35′58″W [2]
Length	70.1 mi (113 km) [3]
Basin	1180 sq mi (3056 km^2) [4]
Discharge	for Lowell, 0.9 miles (1.4 km) from mouth
- average	2822 cu ft/s (80 m^3/s) [5]
- max	35100 cu ft/s (994 m^3/s)
- min	100 cu ft/s (3 m^3/s)

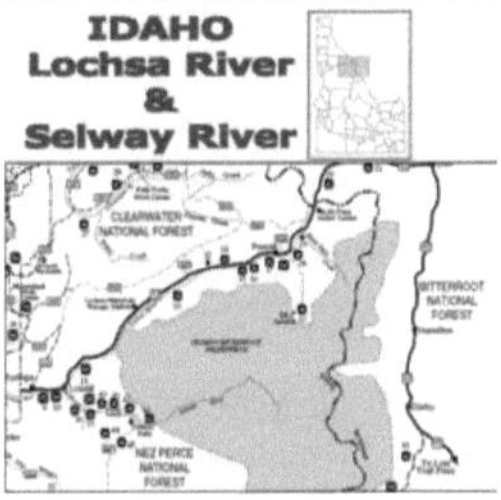

Lochsa and Selway rivers

Location of the mouth of the Lochsa River in Idaho

The **Lochsa River** is located in the northwestern United States, in the mountains of north central Idaho. It is one of two primary tributaries (with the Selway to the south) of the Middle Fork of the Clearwater River in the Clearwater National Forest. *Lochsa* is a Nez Perce word meaning *rough water.*[6] [7]

The Lochsa (pronounced "lock-saw") was included by the U.S. Congress in 1968 as part of the National Wild and Scenic Rivers Act.[8] The Lochsa and Selway rivers and their tributaries have no dams, and their flow is unregulated. In late spring, mid-May to mid-June, the Lochsa River is rated as one of the world's best for continuous whitewater.

The main stem of the Lochsa is 70 miles (113 km) long from its headwaters near Powell Ranger Station in the Bitterroots to Lowell, Idaho, where the Lochsa joins the Selway River to form the Middle Fork of the Clearwater.[3] Over this distance, the river drops nearly 2000 feet (610 m) from 3441 feet (1049 m) above sea level at Powell to 1453 feet (443 m) at Lowell.[2]

The drainage basin for the Lochsa River system covers 1180 square miles (3056 km^2) in Idaho County.[4] The river is fed by the melting of the significant snowpack of the Bitterroot Range.

Course

The Lochsa River begins at the confluence of Crooked Fork and Colt Killed Creek (also called White Sand Creek) near the Powell Ranger Station in northeastern Idaho and flows 70.1 miles (112.8 km) southwest to the small town of Lowell. Running parallel to U.S. Highway 12 for its entire length, the river winds through the Clearwater National Forest in the Bitterroot Mountains. Just below Lowell, the Lochsa and the Selway River join to form the Middle Fork Clearwater River.[9]

At its point of formation just west of Elk Summit Road 360, the Lochsa receives a small tributary, Walton Creek, from the left. The creek flows through a salmon-rearing station before entering the river. Slightly downstream from the salmon station, the Lochsa passes White Sand Campground on the right. About 1.5 miles (2.4 km) later, it flows by the Powell Ranger Station and heliport and, shortly thereafter, Lochsa Lodge and Powell Campground, all on the right, and receives Cliff Creek and Jay Creek, both from the left.[9]

Between 66 miles (106 km) and 65 miles (105 km) from the mouth, the Lochsa receives Papoose Creek from the right, then Robin Creek from the left, and passes Whitehouse Campground and Wendover Campground, both on the right, and receives Wendover Creek from the right. Cold Storage Creek enters from the right at about 64 miles (103 km) from the mouth. Eagle Creek enters from the left about 1 mile (1.6 km) later, and Badger Creek from the right 1 mile (1.6 km) further on. Between 60 miles (97 km) and 50 miles (80 km) from the mouth, the river receives Doe Creek from the right, passes under the Jerry Johnson Pack Bridge, receives Burnt Creek from the right and Warm Springs Creek from the left; passes Jerry Johnson Campground on the right; receives Colgate Creek from the left and Mink Creek on the right; passes Colgate Warm Springs and the Colgate Licks on the right, and receives Bear Mountain Creek from the right, Post Office Creek from the right, Lake Creek from the left, and passes under Mocus Point Pack Bridge, in that order.[9]

From about 48 miles (77 km) to 40 miles (64 km) from the mouth, the Lochsa passes Weir Creek Hot Springs and receives Weir Creek and Ginger Creek, both from the right, Mocus Creek from the left, and Ashpile Creek from the right, then Indian Meadows Creek from the left, Indian Grave Creek from the right, passes the White Pine Access Area on the right, receives Eagle Mountain Creek from the left, and passes under Eagle Mountain Pack Bridge, in that order. Over the next 10 miles (16 km), the river receives Skookum Creek from the right, Stanley Creek from the left, and Castle Creek from the right, passes Green Flat Campground and Nine Mile Rest Area, both on the right, receives Bald Mountain Creek from the right, and Dutch Creek and Hard Creek from the left. Just below 30 miles (48 km) from the mouth, the Lochsa receives Pass Creek from the right and over the next 10 miles (16 km) receives No-See-Um Creek, Dipper Creek, and Sherman Creek, all from the right, then Lone Knob Creek from the left, passes Wilderness Gateway Campground on the left and Boulder Flat on the right, receives Boulder Creek from the left, passes the Lochsa Work Center on the right; receives Sardine Creek, Fish Creek, Bee Creek, Eel Creek, and Otter Slide Creek, all from the right, then passes Beaver Flat on the right and receives Big Stew Creek from the left and Snowshoe Creek from the right.[9]

Walton Creek is the uppermost tributary of the Lochsa River. It flows through a salmon-rearing station before entering the river.

Between about 19 miles (31 km) and 10 miles (16 km) from the mouth, the Lochsa receives Macaroni Creek, Wild Horse Creek, and Tomcat Creek, all from the right, Old Man Creek from the left, Tumble Creek from the right, Split Creek from the left, passes under Split Creek Pack Bridge, receives Fire Creek from the left, Bimerick Creek and Tick Creek from the right, Coolwater Creek from the left, passes Knife Edge Campground and Major Fenn Picnic Area on the right, receives Stub Creek from the left, and Deadman Creek from the right, in that order. Over its last 10 miles (16 km), the river passes Glade Creek Campground and receives Glade Creek, both on the right, passes Apgar Campground and receives Apgar Creek, both on the right, receives Chance Creek from the left, Canyon Creek from the right, Hellgate Creek and Handy Creek, both from the left, Rye Patch Creek from the right, Cat Creek and Kerr Creek, both from the left, Pete King Creek from the right, passes Icicle Spring on the right, receives Lowell Creek from the right, passes a United States Geological Survey (USGS) stream gage 0.9 miles (1.4 km) from the mouth, receives Lottie Creek from the left, passes the town of Lowell on the left, passes under the Lowell Bridge, and meets the Selway River.[9]

Downstream from Lowell, the Middle Fork of the Clearwater River flows west, meets the South and North Forks, and enters the Snake River at Lewiston on the Idaho—Washington border, 98 miles (158 km) by river from Lowell. Below Lewiston, the Snake flows 140 miles (230 km) to its confluence with the Columbia River just south of the Tri-Cities. Over this stretch, four dams in Washington block the Snake. They are the Lower Granite Dam, Little Goose Dam, Lower Monumental Dam, and Ice Harbor Dam, in that order.[9]

History

Long before the arrival of European Americans in the 19th century, the Nez Perce people in Idaho used the trail along the Lochsa River to travel to the plains of Montana to hunt buffalo. Kootenai and Salish (Flatheads) from Montana used the river trail to reach salmon runs in the rivers and streams feeding into the Columbia basin.

During mid-September 1805, the explorers Lewis and Clark traveled westward along the Lolo Trail, descending into and out of the Lochsa Gorge, then above its north rim. They described the region near the modern-day ranger station at Powell as follows:

> " ...the grass entirely eaten out by the horses, we proceeded on 2 miles & Encamped opposit a Small Island at the mouth of a branch on the right side of the river which is at this place 80 yards wide. Swift and Stoney, here we were compelled to kill a Colt for our men & Selves to eat for the want of meat & we named the South fork Colt killed Creek ... The Mountains which we passed to day much worst than yesterday the last excessively bad & thickly Strowed with falling timber & Pine Spruce fur Hackmatak & Tamerack, Steep and Stoney our men and horses much fatigued... "[10]

They experienced significant early season snowfall and suffered near starvation before exiting the mountains onto the Weippe Prairie, where they first encountered the Nez Perce tribe.[11]

In recent history, the U.S. Forest Service has leased areas of the Upper Lochsa for timber extraction. The Forest Service is currently pursuing a controversial land exchange deal with timber and real estate developer Tim Blixseth and Western Pacific Timber.[12] Previous land exchange deals with this developer resulted in one of the largest recorded fines for violating environmental laws.[13] The proposed land swap deal has former forest service senior administrators claiming corruption.[14]

U.S. Route 12

U.S. Route 12 follows the Lochsa River along its north bank.[8] One of the last two-lane U.S. highways constructed, it was completed in the early 1960s, connecting Lewiston with Missoula over Lolo Pass. Two railroads, the Northern Pacific and Union Pacific, had originally planned to ascend the Lochsa; the Northern won the rights and even completed some grades in 1908, but abandoned the project in 1909. The highway along the river was not completed for another 50 years.[15]

Wildlife

Common among the animals using the Lochsa River watershed are deer, elk, black bear on the hillsides and moose in meadows and wet places. Among the birds along the river are Canada Geese, swans, eagles, Osprey and Great Blue Herons. Wild turkeys are seen at lower elevations. More rare are mountain lions, Snow Geese, mountain goats, and Harlequin Ducks. Cutthroat trout, rainbow trout, Chinook salmon and steelhead are among the fish living in the river.[8]

Recreation

Kayaker in Lochsa River

Kayakers and whitewater rafters run the Lochsa, generally between April and June. The "Lochsa River Floating Guide" lists more than 60 rapids between Powell and Lowell, most of which are rated Class III-IV (medium to difficult). Depending on the discharge rate, the level of difficulty can rise in places to Class VI (the limits of navigation) on the International Scale of River Difficulty.[16] The ratings in the "Floating Guide" reflect the difficulty of the Lochsa's rapids when the water levels at the bridge at Lowell are between 5 feet (1.5 m) and 8 feet (2.4 m).[16] The gage at the bridge differs from the USGS gage. A warning to whitewater rafters posted at the USGS web site for its gage says: "This stage reading is from the USGS gaging station 0.7 miles upstream from Lowell bridge, and it may read between 2-3 feet higher than the stage reading on the staff gage at Lowell bridge. This [the USGS gage reading] is not the stage reference to float the Lochsa River."[4] The USGS site advises rafters to check with the Lochsa Ranger District for readings from the bridge.

The "Floating Guide" describes the Class VI rapids at Castle Creek as follows:

> Probably the largest, most technical rapid on the river. A long, twisting rapid whose 1/2-mile length is not totally observable from the water at any one time. Large holes become large waves at high water, and everything pillows left off a huge block of bedrock at the bottom of the rapid. Considered un-runnable at high flows. Must scout. Cannot be seen from highway.[16]

Commercial outfitters offer pre-arranged trips, and people with sufficient technical expertise run the rapids on their own. River access points with parking lots along Highway 12 are at the following sites and highway mile markers: Knife Edge, 107; Split Creek, 111; Fish Creek, 119; Nine Mile, 130, and White Pine, 138.[8]

Other recreation in the watershed includes backcountry skiing, bicycling on part of the TransAmerica Trail or other roads and trails, fly fishing with daily limits or catch-and-release rules, hiking, swimming, camping, and picnicking.

See also

- List of rivers of Idaho

References

[1] Source elevation derived from Google Earth using Geographic Names Information System (GNIS) source coordinates.

[2] "Lochsa River" (http://geonames.usgs.gov/pls/gnispublic/f?p=gnispq:3:::NO::P3_FID:385022). *Geographic Names Information System (GNIS)*. United States Geological Survey (USGS). June 21, 1979. . Retrieved July 26, 2008.

[3] USGS and United States Forest Service. *Rocky Point Quadrangle, Idaho, 7.5-Minute Series (Topographic)* (Map) (1994 ed.).

[4] "USGS 13337000 Lochsa River near Lowell ID" (http://waterdata.usgs.gov/usa/nwis/uv?site_no=13337000). USGS. 2008. . Retrieved July 27, 2008.

[5] The average discharge rate was calculated by adding the average annual discharge rates at the USGS gauge at Lowell for the 77 calendar years, 1930–2006 and dividing by 77.

[6] Space, Ralph S. (2001). *The Lolo Trail, second edition*. Missoula: Historic Montana Publishing. p. 21. ISBN 0-9663355-2-X.

[7] "Lochsa River Canyon" (http://www.lewis-clark.org/content/content-article.asp?ArticleID=2143). *Discovering Lewis & Clark*. VIAs, Inc.. 1998. . Retrieved July 30, 2008.

[8] "Clearwater National Forest: Highway 12: A Long and Winding Road" (http://www.fs.fed.us/r1/clearwater/VisitorInfo/Assets/pdfs/hwy12_corridor.pdf) (pdf). Washington, D.C.: U.S. Dept. of Agriculture, Forest Service. 2004. . Retrieved July 30, 2008.

[9] "Online Topographic Maps from the United States Geological Survey" (http://www.topoquest.com/). TopoQuest. 2008. . Retrieved August 23, 2008.

[10] DeVoto, Bernard (1953). *The Journals of Lewis and Clark*. New York: Houghton Mifflin Company. pp. 238–39. ISBN 0-395-85996-4.

[11] Jackson, Jill (2004). "A Timeline of the Lewis and Clark Expedition" (http://www.lewisandclark.org/?p=exp_timeline&n=landcexp). Lewis and Clark Trail Heritage Foundation. . Retrieved July 30, 2008.

[12] Proposed Upper Lochsa Land Exchange (http://www.fs.fed.us/r1/clearwater/Projects/Upper_Lochsa_LEX/Upper_Lochsa_LEX.htm), US Forest Service, March 2011.

[13] Club owner agrees to record fines for environmental violations (http://www.bozemandailychronicle.com/news/article_c4592fab-2365-55e0-a62c-32c21912e53a.html), Bozeman Chronicle, August 9, 2004.

[14] Retired Forest Service Employees denounce corrupt Upper Lochsa Land Exchange (http://forestpolicy.posterous.com/idaho-retired-forest-service-e), Forest Policy, March 23, 2011.

[15] Conley, Cort (1982). *Idaho for the Curious*. Cambridge, Idaho: Backeddy Books. p. 107. ISBN 0-903566-3-0.

[16] "Lochsa River Floating Guide" (http://www.fs.fed.us/r1/clearwater/VisitorInfo/Assets/pdfs/Lochsa_River_Guide.pdf) (pdf). Northwest Interpretive Association in cooperation with the Clearwater National Forest. 1994. . Retrieved August 24, 2008.

Further reading

Endangered and Threatened Wildlife and Plants; Proposed Designation of Critical Habitat for the Klamath River and Columbia River Distinct Population Segments of Bull Trout (http://www.epa.gov/EPA-SPECIES/2002/November/Day-29/e29232.htm). Washington D.C. ; U.S. Fish and Wildlife Service. Office of Endangered Species, 2002

Hazelbaker, Kris (2004). *Caring for a wild and scenic river* (http://www.fs.fed.us/rm/pubs/rmrs_p034/rmrs_p034_207_216.pdf) In: Shepperd, Wayne D.; Eskew, Lane G., compilers. 2004. *Silviculture in special places: Proceedings of the National Silviculture Workshop; September 8–11, 2003; Granby, CO.* Proceedings RMRS-P-34. Fort Collins, CO: U.S. Department of Agriculture, Forest Service, Rocky Mountain Research Station. pp. 207–16.

Moore, W. R. (1996). *The Lochsa story - Land Ethics In The Bitterroot Mountains*. Missoula, Montana: Mountain Publishing Co. ISBN 0878423419

External links

- Bearpaw River Expeditions (http://www.bearpawexpeditions.com/) - Custom Lochsa Raft Trips
- The Weather Channel (http://epic.weather.com/epicPlayer.html?clip=5987&collection=epic3&nav=141&from=epic&cm_ven=epic&cm_ite=home_maincontent) - Epic Adventures - Lochsa River, Idaho
- Lochsa.info (http://www.lochsa.info/) - stream flows, weather, etc.
- Backcountry Secrets.com (http://backcountrysecrets.com/pointDetails.aspx?pointID=93) - Lochsa River, Idaho
- Idaho Whitewater Trail Map (http://www.idahowhitewatertrail.com/trailmap.aspx) - official state tourism site
- Paddling.net (http://paddling.net/idaho/lochsa.html) - Lochsa River, Idaho
- Topographic map - Lowell, Idaho (http://msrmaps.com/image.aspx?T=2&S=12&Z=11&X=760&Y=6388&W=3&qs=|Lowell|ID|) from USGS via Microsoft Research Maps
- Kayak Idaho.com (http://www.kayakidaho.com/index.php?option=com_g2bridge&view=gallery&Itemid=100192/&g2_itemId=1381&g2_GALLERYSID=81604a10c9594bdd5dcfb8fe3c7819af) - photos - Idaho rivers
- ROW Adventures.com (http://www.rowadventures.com/index.php?page=trips&TripID=25) - Lochsa River
- Rafting Adventures.com (http://www.rafting-adventures.com/lochsariver.htm) - Lochsa River
- Lewis and Clark Trail Adventures (http://www.trailadventures.com/) - Lochsa River
- Lochsa Historical Ranger Station Museum (http://www.visitidaho.org/thingstodo/attractions.aspx?attractionid=31285)

Selway River

The **Selway River** is located in north central Idaho in the northwestern United States within the Selway-Bitterroot Wilderness. It is a tributary of the Middle Fork of the Clearwater River in the Clearwater National Forest. The Selway was one of the rivers included by the United States Congress in 1968 as part of the National Wild and Scenic Rivers Act.

The main stem of the Selway is 100 miles (160 km) in length[1] from the headwaters in the Bitterroots to the confluence with the Lochsa near Lowell to form the Middle Fork of the Clearwater River. The drainage basin for the river system comprises 106977 acres (432.92 km^2) in Idaho County. The river is fed by melting snow out of the mountains of Idaho.

Selway River (Idaho)

Selway River Canyon in Idaho

Courtesy of Idaho Travel Council [1]

Communities

- Lowell

Flora

- Cedar - Western Red Cedar
- ferns
- Firs – Douglas and Grand fir
- Huckleberries
- Pine - Ponderosa Pine
- Spruce - Engelmann Spruce

Wildlife

- Birds
  - Bald Eagle
  - Grouse
  - Heron
  - Osprey
  - Owls
  - Wild Turkey
- Fish
  - Trout – Bullhead, Bull, Rainbow or Steelhead and, Westslope cutthroat trout
  - Salmon- Chinook and Coho
  - Mountain whitefish
- Mammals
  - Bighorn Sheep
  - American Black Bear
  - Cougar
  - Elk
  - Fisher
  - Gray Wolf
  - Lynx
  - Moose
  - Mountain goat
  - Mule deer
  - Northern River Otter
  - Skunk
  - Weasel
  - White-tailed deer
  - Wolverine

Recreation

- Backcountry skiing
- Camping
- Fly fishing – Limits and catch-and-release
- Hiking
- Whitewater rafting – Permits required

References

[1] U.S. Geological Survey. National Hydrography Dataset high-resolution flowline data. The National Map (http://viewer.nationalmap.gov/viewer/), accessed May 3, 2011

Bibliography

- Bugosh, Nicholas. (2000) *Lower Selway River Subbasin assessment.* (http://www.deq.state.id.us/water/data_reports/surface_water/tmdls/selway_river_lower/selway_river_lower_entire.pdf) Lewiston, Idaho : Lewiston Regional Office, Idaho Division of Environmental Quality.
- *Floating the Wild Selway.* (1991) [Missoula, Mont.?] : U.S. Dept. of Agriculture, Forest Service, Northern Region.
- *Selway River Corridor: A Guide to Recreation on the Moose Creek Ranger District* (http://www.fs.fed.us/r1/nezperce/maps/brochures/selwayrivercorridor.pdf). (2000) Kooskia, Idaho : U.S. Dept. of Agriculture, Forest Service, Nez Perce National Forest, Moose Creek Ranger Station.
- *Selway River fisheries investigations : job completion report.* (1979) [Idaho] : Idaho Dept. of Fish & Game.
- *A survey and evaluation of archaeological resources in the Magruder Corridor, Bitterroot National Forest, east-central Idaho, 1969.* (1969) Pocatello, Idaho : Idaho State University Museum.

External links

- NPR Radio expeditions (http://www.npr.org/programs/re/archivesdate/2004/sep/selway/)
- Visit Idaho - Selway River site (http://www.northcentralidaho.info/cgi/TPL05.php?ID=05_999_306)

Clearwater River (Idaho)

Clearwater River	
River	
 Clearwater River near Greer Ferry	
Country	United States
State	Idaho
Regions	Idaho County, Clearwater County, Shoshone County, Nez Perce County
Part of	Snake River
Tributaries	
- left	South Fork Clearwater River
- right	Middle Fork Clearwater River, North Fork Clearwater River, Potlatch River
Cities	Orofino, Lewiston
Primary source	Middle Fork Clearwater River
- location	Confluence of Selway River and Lochsa River, Idaho County
- elevation	1453 ft (443 m)
- coordinates	46°08′28″N 115°35′53″W
Secondary source	South Fork Clearwater River
- location	Near Red River Hot Springs, Idaho County
- elevation	4285 ft (1306 m)
- coordinates	45°52′09″N 115°18′32″W
Source confluence	Kooskia
- location	Idaho County
- elevation	1220 ft (372 m)
- coordinates	46°08′45″N 115°58′56″W
Mouth	Snake River
- location	Lewiston, Nez Perce County
- elevation	741 ft (226 m)

- coordinates	46°25′30″N 117°02′14″W
Length	74.8 mi (120 km), Southeast-northwest [1]
Basin	9645 sq mi (24980 km^2)
Discharge	
- average	15300 cu ft/s (433 m^3/s)
- max	109000 cu ft/s (3087 m^3/s) [2]

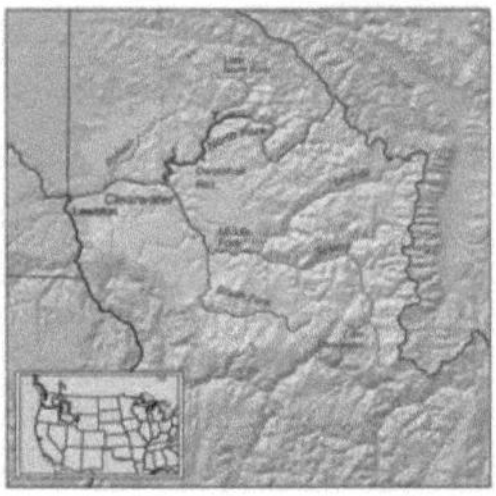

Map showing the Clearwater River watershed

The **Clearwater River** is a 74.8-mile-long (120.4 km)[1] river in north central Idaho, which flows westward from the Bitterroot Mountains along the Idaho-Montana border, and joins the Snake River at Lewiston. In October 1805, the Lewis and Clark Expedition descended the Clearwater River in dugout canoes, putting in at "Canoe Camp," five miles (8 km) downstream from Orofino. By average discharge, the Clearwater River is the largest tributary of the Snake River, although the Salmon River is longer and drains a greater area.

The drainage basin of the Clearwater River is 9645 square miles (24980 km^2). Its mean annual discharge is 15300 cubic feet per second (430 m^3/s)[3]

Course

At the small town of Kooskia, the Middle Fork and South Forks of the Clearwater River join their waters to form the main stem of the Clearwater. The large Middle Fork flows from the Continental Divide in the east, while the much smaller South Fork originates in the Rockies to the south. From the confluence the river meanders northwards, passing a site of the Nez Perce National Historical Park. U.S. Highway 12 follows the river to Kamiah, where it is joined by Lawyer Creek.

The river then continues north through a canyon to the confluence with Lolo Creek. It soon passes the town of Greer and receives Big Creek from the right. At Orofino the river swings northwest and runs in a nearly straight line for about 3 miles (4.8 km), then receives the North Fork Clearwater River from the right at Ahsahka close to Dworshak Reservoir. After the North Fork contributes its flow, the Clearwater trends west and receives Big Canyon Creek from the left, and Bedrock Creek from the right.

As the river canyon cuts deeper into the Columbia Plateau, the Clearwater passes the unincorporated communities of Myrtle and Arrow, where it receives the Potlatch River from the right. Lapwai Creek joins from the left where the river passes close to Spalding. Here, U.S. Highway 95 crosses the Clearwater while Highway 12 continues running along the north shore. The river soon widens and slows into the slack water of Lower Granite Lake as it approaches Lewiston. Just as it crosses the Idaho-Washington state line, it joins its waters with the Snake River.

Tributaries

The Clearwater breaks into several separate forks:

- Clearwater River (west of Orofino to Lewiston-Snake River)
  - Potlatch River (Latah, Clearwater and Nez Perce Counties)
  - **North Fork** Clearwater River (stream, Clearwater County - 46°29′59″N 116°19′49″W; headwaters near Illinois Peak to just west of Orofino)
    - **Little North Fork** Clearwater River (stream, Shoshone & Clearwater Counties; headwaters in south-central Shoshone County, joins the North Fork in the Dworshak Reservoir)
  - **Middle Fork** Clearwater River (stream, Idaho County - 46°08′43″N 115°58′56″W; formed by the confluence of the Selway and Lochsa at Lowell)
  - **South Fork** Clearwater River (stream, Idaho County - 46°08′44″N 115°58′56″W; headwaters near Red River Hot Springs to Kooskia, confluence with the Middle Fork)
    - **Little Clearwater** River (stream, Idaho County - 45°45′11″N 114°46′31″W; near Three Prong Mountain to near Spot Mountain)

River modifications

Clearwater River near Orofino, ID

The Dworshak Reservoir is the only major lake on the Clearwater system, created from the Dworshak Dam, completed in the early 1970s. Dworshak Dam is on the North Fork of the Clearwater River, and is just northwest of Orofino. There is no fish ladder; the dam blocks salmon and steelhead passage.

History

The border between Washington and Idaho was defined as the meridian running north from the confluence of the Clearwater River and the Snake River. Although this border is often referred to as the 117th meridian west longitude, the actual border line is slightly west (less than 2 miles) of the 117th meridian.[4]

References

[1] U.S. Geological Survey. National Hydrography Dataset high-resolution flowline data. The National Map (http://viewer.nationalmap.gov/viewer/), accessed May 3, 2011

[2] "USGS Gage #13343000 on the Clearwater River near Lewiston" (http://nwis.waterdata.usgs.gov/nwis/peak?site_no=13343000&agency_cd=USGS&format=html). *National Water Information System*. United States Geological Survey. . Retrieved 2010-03-09.

[3] Clearwater Subbasin Plan (http://www.nwcouncil.org/fw/subbasinplanning/clearwater/plan/assessment.pdf), Northwest Power and Conservation Council

[4] Washington State Constitution (http://www1.leg.wa.gov/LawsAndAgencyRules/Constitution.htm) Arcticle XXIV Boundaries

External links

- U.S. Geological Survey Geographic Names Information System: Clearwater River (http://geonames.usgs.gov/pls/gnispublic/f?p=gnispq:3:::NO::P3_FID:379406), USGS GNIS

Trapper Peak (Montana)

Trapper Peak	
 Trapper Peak, as seen from the Boulder Point Lookout	
Elevation	10157 ft (3096 m) [1]
Prominence	3570 ft (1088 m) [1]
Location	
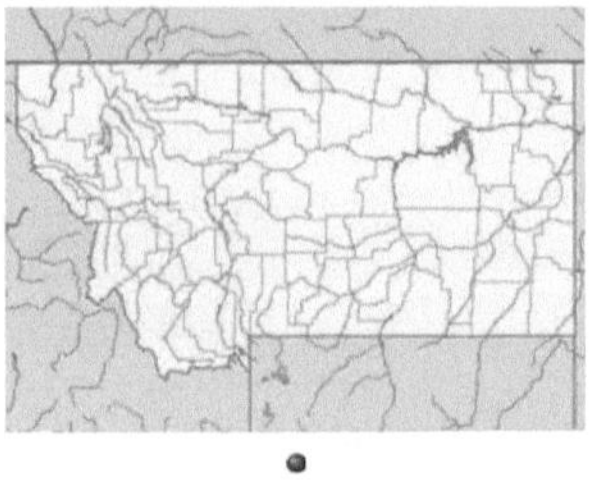 Trapper Peak	
Range	Bitterroot Mountains
Coordinates	45°53′24″N 114°17′51″W[2]
Topo map	USGS Trapper Peak, MT
Climbing	
Easiest route	Hike

Trapper Peak is the highest point in the Bitterroot Mountains, part of the larger Bitterroot Range in western Montana. The peak is located within the Central Bitterroot subrange of the Bitterroot Mountains and the Selway-Bitterroot Wilderness Area. It rises over 6000 feet (1830 m) above the nearby Bitterroot Valley. A trail to the peak climbs 3800 feet (1160 m) from the end of a Forest Service road.

References

[1] "Trapper Peak" (http://www.bivouac.com/MtnPg.asp?MtnId=10414). Bivouac.com. . Retrieved 2008-12-20.

[2] "Trapper Peak" (http://geonames.usgs.gov/pls/gnispublic/f?p=gnispq:3:::NO::P3_FID:792352). Geographic Names Information System, U.S. Geological Survey. .

External links

- "Trapper Peak" (http://www.summitpost.org/page/353250). SummitPost.org.
- "Trapper Peak, Montana" (http://www.peakbagger.com/peak.aspx?pid=4983). Peakbagger.com.

Trapper Peak from US 93 near Conner

Ravalli County, Montana

Ravalli County, Montana

Location in the state of Montana

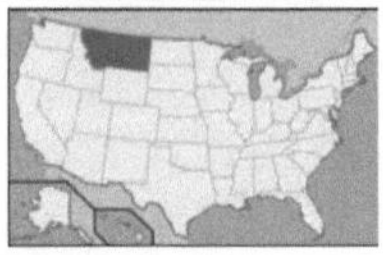

Montana's location in the U.S.

Founded	1893
Seat	Hamilton
Largest city	Hamilton
Area **- Total** **- Land** **- Water**	2400 sq mi (6216 km²) 2394 sq mi (6200 km²) 6 sq mi (16 km²), 0.25%
Population **- (2010)** **- Density**	40212 16/sq mi (6/km²)
Time zone	Mountain: UTC-7/-6
Website	www.co.ravalli.mt.us [1]

Ravalli County is a county located in the southwestern part of the U.S. state of Montana. Ravalli County is part of a north/south mountain valley bordered by the Sapphire Mountains on the East and the Bitterroot Mountains on the West. It is often referred to as the Bitterroot Valley and is named for the Bitterroot Flower. As of 2010, the population was 40,212. Its county seat is Hamilton[2] .

The county is entirely on the Pacific Ocean side of the Continental Divide, which follows the Idaho-Montana border from Wyoming until Ravalli County. Here it turns east into Montana, between Chief Joseph Pass and Lost Trail Pass, and follows the Ravalli County-Beaverhead County border.

History

Ravalli County was once home to the Bitterroot Salish tribe. The tribe was first encountered in 1805 by the Lewis and Clark Expedition, which noted the friendly nature of the tribe. The Catholic Church took interest in creating a mission in the area, and in 1841, Stevensville was founded. In 1891, the Salish tribe was relocated to the current Flathead Reservation under the Treaty of Hellgate.

In 1877, Chief Joseph and his Wallowa band of Nez Perce passed through Ravalli County in their attempt to escape confinement to a reservation; they were captured en route to Canada just south of Havre.

Ravalli County was created by the Montana Legislature on March 3, 1893. It is named after the Italian Jesuit priest Antony Ravalli, who arrived in the Bitterroot Valley in 1845.

Economy

The historical economy of Ravalli County lies predominantly in agriculture and timber. Marcus Daly, one of three Butte copper kings, funded logging operations in the Bitterroot Valley. The lumber was necessary for the Butte copper operation. Recently, more of Ravalli County's economy stems from tourism. The valley borders the Selway-Bitterroot Wilderness and offers a wide variety of wildlife, including some of the few remaining wolverine and wolf populations in the contiguous states. The Lost Trail Powder Mountain ski area is at Lost Trail Pass on the Idaho border on US-93.

Geography

According to the United States Census Bureau, the county has a total area of **unknown operator: u','** square miles (**unknown operator: u'strong'unknown operator: u','**km^2), of which, **unknown operator: u','** square miles (**unknown operator: u'strong'unknown operator: u','**km^2) of it is land and 6 square miles (16 km^2) of it (0.25%) is water.

Major highways

- 93 U.S. Highway 93
- 38 Montana Highway 38

Adjacent counties

- Missoula County - north
- Granite County - east
- Beaverhead - southeast
- Lemhi County, Idaho - south
- Idaho County, Idaho - west

National protected areas

- Bitterroot National Forest (part)
- Lee Metcalf National Wildlife Refuge
- Lolo National Forest (part)

Demographics

Historical populations

Census	Pop.	%±
1900	7822	—
1910	11666	49.1%
1920	10098	−13.4%
1930	10315	2.1%
1940	12478	21.0%
1950	13101	5.0%
1960	12341	−5.8%
1970	14409	16.8%

1980	22493		56.1%
1990	25010		11.2%
2000	36070		44.2%
2010	40212		11.5%
sources:[3] [4]			

As of the census of 2000 [5] , there were 36,070 people, 14,289 households, and 10,188 families residing in the county. The population density was 15 people per square mile (6/km²). There were 15,946 housing units at an average density of 7 per square mile (3/km²). The racial makeup of the county was 96.71% White, 0.14% Black or African American, 0.88% Native American, 0.30% Asian, 0.10% Pacific Islander, 0.44% from other races, and 1.44% from two or more races. 1.88% of the population were Hispanic or Latino of any race. 22.1% were of German, 14.1% English, 11.1% Irish, 7.9% American and 6.3% Norwegian ancestry according to Census 2000.

There were 14,289 households out of which 30.20% had children under the age of 18 living with them, 60.30% were married couples living together, 7.50% had a female householder with no husband present, and 28.70% were non-families. 24.10% of all households were made up of individuals and 9.80% had someone living alone who was 65 years of age or older. The average household size was 2.48 and the average family size was 2.94.

In the county the population was spread out with 25.60% under the age of 18, 6.20% from 18 to 24, 24.70% from 25 to 44, 28.00% from 45 to 64, and 15.50% who were 65 years of age or older. The median age was 41 years. For every 100 females there were 98.60 males. For every 100 females age 18 and over, there were 95.40 males.

The median income for a household in the county was $31,992, and the median income for a family was $38,397. Males had a median income of $30,994 versus $19,987 for females. The per capita income for the county was $17,935. About 9.60% of families and 13.80% of the population were below the poverty line, including 20.10% of those under age 18 and 6.30% of those age 65 or over.

Communities

City

- Hamilton

Towns

- Darby
- Pinesdale
- Stevensville

Census-designated places

- Corvallis
- Florence
- Victor

Notable officials

- Henry L. Myers, later a United States Senator from Montana, was prosecuting attorney for Ravalli County from 1895 to 1899.

See also

- List of cemeteries in Ravalli County, Montana
- List of lakes in Ravalli County, Montana
- List of mountains in Ravalli County, Montana
- National Register of Historic Places listings in Ravalli County, Montana

References

[1] http://www.co.ravalli.mt.us
[2] "Find a County" (http://www.naco.org/Counties/Pages/FindACounty.aspx). National Association of Counties. . Retrieved 2011-06-07.
[3] census.gov - Montana population by county, 1900-90 (http://www.census.gov/population/cencounts/mt190090.txt) - accessed 2009-05-02
[4] quickfacts.census.gov - Ravalli County (http://quickfacts.census.gov/qfd/states/30/30081.html) - accessed 2009-05-02
[5] "American FactFinder" (http://factfinder.census.gov). United States Census Bureau. . Retrieved 2008-01-31.

External links

- Ravalli County and the Bitterroot Valley websites (http://www.bitterrootsites.com)

Missoula County, Montana

Missoula County, Montana

Location in the state of Montana

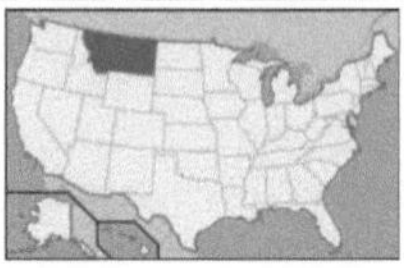

Montana's location in the U.S.

Founded	1864
Seat	Missoula
Largest city	Missoula
Area **- Total** **- Land** **- Water**	2618 sq mi (6781 km²) 2598 sq mi (6729 km²) 20 sq mi (52 km²), 0.78%
Population **- (2010)** **- Density**	109299 36/sq mi (14/km²)
Website	www.co.missoula.mt.us [1]

Missoula County is a county located in the U.S. state of Montana, with Missoula as its county seat and largest city[2]. It is the only county in the Missoula Metropolitan Area. The 2010 census put the Missoula County population at 109,299 making it the second most populous county in Montana.

Geography

According to the United States Census Bureau, Missoula County has a total area of **unknown operator: u','** square miles (**unknown operator: u'strong'unknown operator: u','**km^2), of which, **unknown operator: u','** square miles (**unknown operator: u'strong'unknown operator: u','**km^2) of it is land and 20 square miles (52 km^2) of it (0.78%) is water. Five large valleys and two major rivers wind through this mountainous region.

Major highways

- 90 Interstate 90
- 12 U.S. Highway 12
- 93 U.S. Highway 93
- 83 Montana Highway 83
- 200 Montana Highway 200

Adjacent counties

- Mineral County - west
- Sanders County - northwest
- Lake County - north
- Flathead County - northeast
- Powell County - east
- Granite County - southeast
- Ravalli County - south
- Idaho County, Idaho - southwest
- Clearwater County, Idaho - southwest

National protected areas

- Bitterroot National Forest (part)
- Flathead National Forest (part)
- Lolo National Forest (part)
- Rattlesnake National Recreation Area

Demographics

Historical populations		
Census	**Pop.**	**%±**
1870	2554	—
1880	2537	−0.7%
1890	14427	468.7%
1900	13964	−3.2%
1910	23596	69.0%
1920	24041	1.9%
1930	21782	−9.4%
1940	29038	33.3%
1950	35493	22.2%
1960	44663	25.8%
1970	58263	30.5%
1980	76016	30.5%
1990	78687	3.5%
2000	95802	21.8%
2010	109299	14.1%
sources:[3] [4] [5]		

As of the 2000 census[6] , there were 95,802 people, 38,439 households, and 23,140 families residing in Missoula County. The population density was 37 people per square mile (14/km²). There were 41,319 housing units at an average density of 16 per square mile (6/km²). The racial makeup of the county was:

- 94.02% White

- 0.27% Black or African American
- 2.29% Native American
- 1.02% Asian
- 0.08% Pacific Islander
- 0.45% from other races
- 1.86% from two or more races.

1.61% of the population were Hispanic or Latino of any race. 22.7% were of German, 13.0% Irish, 10.4% English, 8.5% Norwegian and 5.6% American ancestry according to Census 2000.

There were 38,439 households out of which 29.20% had children under the age of 18 living with them, 47.40% were married couples living together, 9.20% had a female householder with no husband present, and 39.80% were non-families. 28.00% of all households were made up of individuals and 7.50% had someone living alone who was 65 years of age or older. The average household size was 2.40 and the average family size was 2.96.

In the county the age distribution of the population shows 22.90% under the age of 18, 15.40% from 18 to 24, 29.20% from 25 to 44, 22.60% from 45 to 64, and 10.00% who were 65 years of age or older. The median age was 33 years. For every 100 females there were 99.90 males. For every 100 females age 18 and over, there were 98.00 males.

The median income for a household in the county was $34,454, and the median income for a family was $44,865. Males had a median income of $31,605 versus $21,720 for females. The per capita income for the county was $17,808. About 8.80% of families and 14.80% of the population were below the poverty line, including 14.60% of those under age 18 and 8.20% of those age 65 or over.

Communities

City

- Missoula

Census-designated places

- Bonner-West Riverside
- Clinton
- East Missoula
- Evaro
- Frenchtown
- Lolo
- Orchard Homes
- Seeley Lake
- Wye

Other community

- Condon

History

Missoula County, Washington Territory was incorporated in 1860. Missoula County encompassed present-day Missoula and Deer Lodge Counties, as well as a good portion of land north and south of present-day Missoula County. Hell Gate Town, the county seat, was located at the confluence of the Clark Fork and Bitterroot Rivers.

In 1863, Missoula County reverted to the Idaho Territory when it was created in that year. Missoula County then shifted its boundaries to look more like its present-day boundaries dividing the Bitterroot Mountains from the Rocky Mountains. Missoula County became a part of Montana Territory when the territory was organized out of the existing Idaho Territory by Act of Congress and signed into law by President Abraham Lincoln on May 26, 1864. This early Missoula County was one of the eleven original counties of Montana Territory and consisted of part or all of the lands now occupied by the current Ravalli, Missoula, Granite, Mineral, Lake, Sanders, Lincoln, Flathead, and Glacier Counties.

Notable residents

- Marc Racicot, former Governor of Montana, previously served as deputy county attorney here.
- Diane Sands, member of the Montana House of Representatives
- Edgar Samuel Paxson, frontier artist of the Old West

See also

- List of cemeteries in Missoula County, Montana
- List of lakes in Missoula County, Montana
- List of mountains in Missoula County, Montana
- National Register of Historic Places listings in Missoula County, Montana

References

[1] http://www.co.missoula.mt.us
[2] "Find a County" (http://www.naco.org/Counties/Pages/FindACounty.aspx). National Association of Counties. . Retrieved 2011-06-07.
[3] census.gov Montana population by county, 1900-90 (http://www.census.gov/population/cencounts/mt190090.txt) - accessed 2009-05-02
[4] quickfacts.census.gov - Missoula County (http://quickfacts.census.gov/qfd/states/30/30063.html) - accessed 2009-05-02
[5] "Missoula County QuickFacts from the US Census Bureau" (http://quickfacts.census.gov/qfd/states/30/30063.html). Quickfacts.census.gov. . Retrieved 2011-11-14.
[6] "American FactFinder" (http://factfinder.census.gov). United States Census Bureau. . Retrieved 2008-01-31.

External links

- Missoula County official website (http://www.co.missoula.mt.us/)

Article Sources and Contributors

Selway-Bitterroot Wilderness *Source*: http://en.wikipedia.org/w/index.php?title=Selway-Bitterroot_Wilderness *Contributors*: Axcordion, Backspace, BitterrootBob, Bobblewik, BrokenSphere, Cdc, Clarkbhm, Droll, Dsetay, G. Thomas, Gene Nygaard, Glacier109, GrahamHardy, Hike395, Hmains, Jengod, Jllm06, Loren.wilton, MONGO, Robbiegiles, Ryanniemi, Soap, Staffelde, 3 anonymous edits

Strict Nature Reserve/Wilderness Area *Source*: http://en.wikipedia.org/w/index.php?title=Strict_Nature_Reserve/Wilderness_Area *Contributors*: Bermicourt, BlueCanoe

Idaho *Source*: http://en.wikipedia.org/w/index.php?title=Idaho *Contributors*: -Kerplunk-, 16@r, Abemedina, Academic Challenger, Acoma DC, Acroterion, Adamciks, Adrian.benko, Aesopos, Aeusoes1, Aflafla1, AgentPeppermint, Ahassan05, Ahoerstemeier, Ahuskay, Aivazovsky, Alai, Alaniaris, Alansohn, Albrozdude, Alex43223, AlexiusHoratius, Alexwcovington, Alinawazadil, Alisproparis, Allen4names, Allmightyduck, Allstarecho, Altenmann, Am wikicite, Amalthea, Amazonien, Amikeco, Andonic, Andrew Kelly, Andrewlp1991, Andrewspy007, Andspd1980, Andy Marchbanks, Andy120290, AndyZ, Angbrown, Angr, Animum, Anlala, Aram33, Archtransit, ArglebargleIV, Arrabbikum, Asidemes, Ask123, Astynax, Atomicskier, AtticusX, AuburnPilot, Aude, Avecchione, Avono, AzaToth, Aztom2, BSveen, Backslash Forwardslash, Barek, Bazonka, BazookaJoe, Bbarney, Beano, Beansbaxter, Beland, BenBaker, BenH, Bender235, Benjh40, Bentley4, Bettia, Bigcheesebebbs, BigrTex, Bill Price 41, Blackhobo23, Blanchardb, Blastingcap, BlueAzure, BlueMoonlet, Blueboy96, Bluedogtn, Blueflashlight07, Blur127, Bobblewik, Boing! said Zebedee, BoiseRedHead, Boisemedia, Bomac, Bongwarrior, Bonzostar, BookDude68, Bookssaan, BorgHunter, Bourquie, Braaropolis, Brad410, Bremerenator, Brennan626, Brequinda, BrianGV, Brianga, Brion VIBBER, Brothejr, Btball, Buaidh, Butwhatdoiknow, Byronshock, CART fan, CDA, CJLL Wright, CWii, Cadiebug94, CalicoCatLover, Caltas, Calton, Calvin 1998, Camcgrath, Camw, Can't sleep, clown will eat me, CanisRufus, Capnganch, Caponer, Capricorn42, Captain panda, Carlaude, CasperBraske, Catbtds, CattleGirl, Cburnett, Cchow2, Centrx, Chanheigeorge, Charlesaaronthompson, ChaseDave, Chasedawg1, Chlo nb, Chowells, ChrisRuvolo, Chrislk02, Chrisn4255, Chun-hian, Circumspect, Civil Engineer III, ClairSamoht, Clariline, Clarkbhm, Closedmouth, Cocoaguy, Coemgenus, CommonsDelinker, Conscious, Conversion script, Cool3, CopperSquare, Cormac Canales, Coulraphobic123, Courcelles, Craiguk4, Cremepuff222, Cribbswh, Curps, Cutiepie41698, D3R3CK, D6, DCEdwards1966, DFS454, DMacks, DRTllbrg, Dabomb87, Dale Arnett, Damicatz, DanielCD, Danny, Danny7814, Darth Panda, Daven200520, David Justin, DavidOaks, Dbtfz, Dcornwall, Dedee73, Deflective, Demi, Denniss, Deor, DerHexer, Diamondback2222, Difu Wu, Discospinster, Dividing, DivineIntervention, Djackino, Dodders, Dominus, Dp462090, Dparvin, Dr. Blofeld, Dragon, Dralwik, Dreadstar, Drilnoth, Drumguy8800, Drumlineramos, Dthomsen8, Dustinr88, Dxco, Dysmorodrepanis, ESkog, Eastlaw, Ebguy, Eco84, Ed g2s, Edward, Eheaney98, El C, Elassint, Elockid, Emeraldcityserendipity, Emhoo, Emurphy42, EncycloPetey, EnemyOfTheState, EnglishEfternamnn, Entirelybs, EoGuy, Epbr123, Epolk, Erebus Morgaine, Erik9, Est300, EstebanF, Eu.stefan, EvanSvensson, Everyking, Evice, Excirial, Extransit, Faradayplank, Faustus37, Favonian, Fbarabas, FeanorStar7, Felt, Fishing, Flatterworld, Fnlayson, FoggieLlama, Fogherty V. Tatin, Footwarrior, Fram, Fredlyfish4, Freemarket, Frietjes, FriscoKnight, Funnyhat, Fw208, GB fan, GWhitewood, Gadfium, Gaff, Gaius Cornelius, Gall1314, Gatewaycat, Gdr, Geo.plrd, Geobeedude, Gerrish, Ghaly, Gil Gamesh, Gilliam, Gingerar, Ginqui, Glacier109, Gnadenhuten, Gobonobo, Gogo Dodo, Gogrimm, Good Olfactory, Goodnightmush, Graham87, Grandpafootsoldier, GregU, Grinned2death, Gscshoyru, Guiltlessgecko, Guoguo12, Gurch, Gus831, Góngora, H-stt, HJ Mitchell, Habes1234, Hagerman, Halo808, Hayden120, Hdt83, Heegoop, Heracles31, Hibernian, Highvale, Hires an editor, Hitssquad, Hmains, Hojimachong, Horologium, Hugh7, Humpadoor, Hushpuckena, Husond, Hut 8.5, IBook of the Revolution, Iamidaho, Idaho College Office, Idaho spud, Idahoan, Idahoancrow, Idahopotatoe, Idawho, Ilikegrls2043, Ilo333, Immunize, Indefatigable, Infoporfin, Interiot, Italiasian, Itcousin, Ivirivi00, Ivy WASC, Ixitol, J.Gowers, J.delanoy, JCNelson, JDBush, JDoorjam, JFreeman, Jack Cox, Jack Henderson, Jack Wills It, Jacob.jose, Jaeger5432, Jafeluv, JamesBWatson, Jamesgarfield1022, Jamesontai, Jared Preston, Jason Leach, Jauhienij, Jcam, Jeff G., Jeffq, Jehorn, Jengod, Jeremiestrother, Jeremybrooks, JesseBHolmes, Jfknrh, Jhendin, Jim McKeeth, JimIrwin, JinJian, Jni, Joan53, Joao Xavier, JobSmee, Jockyinfront, JoeSmack, John K, JohnInDC, JohnWittle, Jojhutton, Jon Stockton, Jonel, Jordgettte, Jose77, Joshuamills74, Juzeris, Jwamick, KJS77, Kablammo, Kahuroa, Kaisershatner, Karthikc123, Kazrak, Kbh3rd, Keilana, Keith Edkins, KelvinSmith, Ken g6, Kerryboy1, Kfengler, Khatru2, Khazar, Khip44, King8954220064846845, Kintetsubuffalo, Kj24, Knight of BAAWA, KnowledgeOfSelf, Kokilio, Kozuch, Kralizec!, Krich, Kristinpedia, Kubigula, Kukini, Kumioko, Kungming2, Kuru, Kwamikagami, Kyle1278, L Kensington, Larsbars, Ld, LeaveSleaves, Levil, Levineps, LifeStar, Lightmouse, Lilac Soul, Limideen, Linesdity, Linuxerist, Litefantastic, Log 10000 = 4, Longhair, Longnskinny, Looxix, Lord Voldemort, Lorty, LostBurner, Lostpl8, Ltwin, LukeM212, Lulu79151, Luna Santin, M C Y 1008, M.nelson, MBob, MC MasterChef, MER-C, MJCdetroit, MacGyverMagic, MacTire02, Magioladitis, Makaristos, Malepheasant, Malpass93, Marek69, MarkSweep, Matticus78, Mattster395, Mav, May1995, Mcghiever, Mcmcs, Mdk1998, Melvalevis, Mentifisto, Michaelmas1957, Mike 7, Mike s, Mike4ty4, MikeLynch, Mikevegas40, Millahnna, Minesweeper, Minna Sora no Shita, Mintleaf, Mirodoc, MisterSheik, Mizzizzatt, Mkilly, Mmdoogie, Monterey Bay, Morgangrether, Moriori, Moulder, Mountain walrus, Moverton, Moyabrit, Mr. Lefty, Mrhyak, Mrjohnhancock, Mu Cow, Mwanner, Mxn, Myanw, Mübarek-Demokrasi, NE2, NEICenergy, Nakon, Nalleteddy, NapoliRoma, Nareek, Natesvp, Nathan moonhead, NatureA16, NatusRoma, Neurolysis, Nevermore27, Neverquick, NewEnglandYankee, Nice2cu09, Nick in syd, Niklem, Niteowlneils, Njoedits, Nkrosse, NorCalHistory, Northwesterner1, Nothingofwater, Nposs, NuclearWarfare, Nufy8, Nvog86, Nyttend, Ohconfucius, Ohnoitsjamie, OlEnglish, Oleg Alexandrov, Omicronpersei8, One, Onesimus73, Osloboslo, Ostephy, OverlordQ, P4k, PL290, Pampas Cat, Pan Dan, Paris1127, Paulc206, Pavel Vozenilek, PdxCobra, Pepsidrinka, Peter McGinley, Peytonbland, Pfly, Pgk, Philthecow, Philwelch, PhnomPencil, Picklegnome, Pilotguy, Pinkadelica, Pippu d'Angelo, Pitamakan, Pkeck, Pl2241, Poconnor1984, Pol098, Pol430, Possum, Post Falls Man, Postdlf, Prattflora, PrincessofLlyr, Prolog, PseudoSudo, Psy guy, Puget Sound, QQQ, Qqqqqq, QuackyQuackDuck, Quasipalm, Queson, R.T.Gellar, RL0919, RPIRED, Ra2319, Raul654, Ravedave, Raven in Orbit, Ray Van De Walker, Raymondwinn, Reach Out to the Truth, Realm of Shadows, Reaper Eternal, Reconfirmer, RedRollerskate, Reedy, Reesh, Republic of North Idaho, Retinarow, Reztip, Rich Farmbrough, Rich257, Richard D. LeCour, Rick Block, RicoRichmond, Rjust, Rjwilmsi, Rkmlai, Rmhermen, Rob Hooft, Rob Salzman, RobLa, Robbie098, Robbiegiles, Robertb-dc, Robina Fox, Robomaeyhem, Rocastelo, Rockford1963, Rockford63, Roffle23b, Rohdeaa, Roke, Romanm, Ron2, Ronhjones, Ropers, Rorschach, Ross Burgess, Rothorpe, Rrburke, Rreagan007, Rschmertz, Ruhrfisch, Ruy Pugliesi, Ryulong, SD6-Agent, SHIMONSHA, SNIyer12, SPUI, STRH008, Saculsacul, SailorMoon007, Sam0321, Samuell, Sancho89, Saopaulo1, Sarasuex7, SarekOfVulcan, Satori Son, Saturation2, SatyrTN, Sben87, Sca, Scanlan, Scarab12, Schaengel89, SchfiftyThree, Scott Burley, Seaphoto, Seb az86556, Seicer, Serkul, Sesel, Sfmontyo, Shadowjams, Shadzane, Shanes, Shotgun2197, Shwheeler, Silly rabbit, Sjö, Sk8ski, Skyezx, Skylap, Slant6guy, Slon02, Smith03, SmthManly, Snowdog, Socalpotato, Solipsist, Solutus, SoulMeetsBody, Sparky the Seventh Chaos, Spartan397, Spencer, Spencer F., Spin2219, SpuriousQ, Spydrlink, Squid661, Stanistani, Staplegunther, Stepp-Wulf, Stuartkonen, Sumoeagle179, Sven Manguard, Swollib, Swoods83, TOttenville8, TUF-KAT, Tabletop, Tach47, Tawker, Tcncv, Teak the Kiwi, Template namespace initialisation script, Tensai, Tesscass, Thankyouforsmokingplease, The Epopt, The Lizard Wizard, The Obento Musubi, The Realms of Gold, The Singing Badger, The Thing That Should Not Be, The Transhumanist, The Utahraptor, The undertow, TheProject, Thehelpfulone, Thejohnnyutah, Thelockedown, Theroadislong, TheronWare, Thingg, Thiseye, Thrind, Thuylinhpham, Tide rolls, Tidying Up, Timneu22, Tinosa, Tmg13, Tmopkisn, Tmtresh, Tobby72, Toiletpaper77, Tommy2010, Tpbradbury, TreyL3, Trimalchio, Triona, Trotter, Trubye, Trusilver, Twas Now, Twi20Pi, TwinCityIL, Tygrrr, Tyrol69, Ufwuct, Ugur Basak, Uncle Dick, UnitedStatesian, Unixan, Unschool, Usseoul, Valfontis, Vandalfan04, Vary, Vcap51, Velvetron, Versus22, Viriditas, Vishnava, Vpuliva, Vsmith, Vspencer, Vzbs34, W.stanovsky, WPANI, Waliy sherpa, Walton One, Wangi, Wapcaplet, Wars, WarthogDemon, Wayne Slam, Wayward, Weatherman78, Weetoddid, Welsh, Wereon, Wfaulk, Wgbc2032, Wheelz45, WhisperToMe, WhiteCat, Why Not A Duck, Wikisidd, William Avery, Winged Cat, Wintonian, Wmahan, X!, X96lee15, Xiner, Xxxiv34, Xyzzy n, Xyzzyva, Yamamoto Ichiro, Yboi, YellowMonkey, Youngamerican, Yserarau, Yuckfoo, Yurik, Yuuzhan, Z.E.R.O., Zafiroblue05, Zdv, Zenohockey, Zereshk, Zhernovoi, Zits24, Zoe, ZooFari, Zscout370, Zson, Zzedar, Zzyzx11, Δ, , 1615 anonymous edits

Bitterroot Range *Source*: http://en.wikipedia.org/w/index.php?title=Bitterroot_Range *Contributors*: Allen3, BD2412, BitterrootBob, BrownHairedGirl, Bumm13, Carders, Crushin2000, Decumanus, Douzzer, Droll, Drumguy8800, Eric-Wester, Finn-Zoltan, G. Thomas, Geologyguy, Grutness, Hmains, JHunterJ, JIP, Jblykins, Keilana, Lightmouse, Look2See1, Mhockey, Mindmatrix, Pmc, RJFJR, RedWolf, ScaldingHotSoup, Soap, Stan Shebs, Waltpohl, 22 anonymous edits

Bitterroot National Forest *Source*: http://en.wikipedia.org/w/index.php?title=Bitterroot_National_Forest *Contributors*: Acroterion, Backspace, Best O Fortuna, BitterrootBob, Bookgrrl, BrokenSphere, Clarkbhm, Cornbreadearl, Daman93, Darwinek, Droll, Drumguy8800, Dsetay, El C, Figma, GrahamHardy, Hmains, Jllm06, Kindkit, Kwamikagami, Lightmouse, MONGO, Mike Cline, Nightscream, Plumu2010, Raven4x4x, Rjensen, Rjwilmsi, Soap, Tangurena, Ulric1313, Utopies, Wavelength, 7 anonymous edits

Clearwater National Forest *Source*: http://en.wikipedia.org/w/index.php?title=Clearwater_National_Forest *Contributors*: Acroterion, Allen3, Backspace, Bobblewik, BrownHairedGirl, Clarkbhm, Cornbreadearl, Dawynn, Dontexpect, Droll, DuncanHill, Fingers-of-Pyrex, GrahamHardy, Grutness, Jengod, Jllm06, Jtranger, Lightmouse, MONGO, PDH, Plasticup, Rmhermen, Robbiegiles, Sluzzelin, SpookyPig, Tangurena, 4 anonymous edits

Lolo National Forest *Source*: http://en.wikipedia.org/w/index.php?title=Lolo_National_Forest *Contributors*: Acroterion, Backspace, Clarkbhm, Cornbreadearl, Darwinek, Droll, Hauganm, Jllm06, Kwamikagami, Lantay77, Lightmouse, Lol pop chicken, Luckyz, MONGO, Mike Cline, Northfork Mike, Plasticup, Simon12, Skeezix1000, Skookum1, Tangurena, Utopies, 1 anonymous edits

Frank Church—River of No Return Wilderness *Source*: http://en.wikipedia.org/w/index.php?title=Frank_Church%E2%80%94River_of_No_Return_Wilderness *Contributors*: Accesspig, Anthony22, Axcordion, Backspace, Binarypower, Bobblewik, Cdc, Clarkbhm, Droll, Drumguy8800, FastLizard4, Fingers-of-Pyrex, Fixbeforeigo, Gimie the beat boys, GrahamHardy, Ground Zero, Hike395, Hmains, Jengod, Kaszeta, KirbyMeister, Lars Washington, Lightmouse, MONGO, Mountainlegends, Pauli133, Preslethe, Reywas92, Rjwilmsi, Ryanniemi, Sharalds, Synergy, YUL89YYZ, Yllosubmarine, 16 anonymous edits

Nez Perce National Forest *Source*: http://en.wikipedia.org/w/index.php?title=Nez_Perce_National_Forest *Contributors*: A2Kafir, Acroterion, Backspace, Clarkbhm, Clawed, Cornbreadearl, Dawynn, Droll, GrahamHardy, Grutness, Hmains, Jengod, Jllm06, Kwamikagami, Lightmouse, MONGO, Robbiegiles, Sonett72, Spangineer, Tangurena, Zoicon5, 3 anonymous edits

Lochsa River *Source*: http://en.wikipedia.org/w/index.php?title=Lochsa_River *Contributors*: Bender235, D6, David stokar, Drumguy8800, Finetooth, Glacier109, Hmains, Ipoellet, Kacbal, Kbreezy04, Ken Gallager, Kwamikagami, MONGO, Philkon, PhnomPencil, R'n'B, Robbiegiles, SDC, Skapur, Star Mississippi, Trnsproducers, UtherSRG, Zyxw, 11 anonymous edits

Selway River *Source*: http://en.wikipedia.org/w/index.php?title=Selway_River *Contributors*: Glacier109, Hmains, J.delanoy, Ken Gallager, Lightmouse, MONGO, Mcguy0, Nposs, Robbiegiles, UtherSRG, 4 anonymous edits

Clearwater River (Idaho) *Source*: http://en.wikipedia.org/w/index.php?title=Clearwater_River_%28Idaho%29 *Contributors*: Alan Liefting, AllyUnion, Bazonka, Big Bob the Finder, Chris the speller, Conversion script, D6, Decumanus, Delirium, Dsdugan, Editore99, Fingers-of-Pyrex, George Burgess, Glacier109, Graham87, Grutness, Hmains, Howcheng, J.delanoy, JustAGal, Ken Gallager, Kmusser, Lightmouse, Lukobe, MONGO, Meister, Pfly, R'n'B, RedWolf, Rich Farmbrough, Rkmlai, Robbiegiles, Shannon1, Smartone100, Steinsky, Stepheng3, TexasAndroid, WikiDon, 24 anonymous edits

Trapper Peak (Montana) *Source*: http://en.wikipedia.org/w/index.php?title=Trapper_Peak_%28Montana%29 *Contributors*: Backspace, Bluemoose, Carders, Droll, Drumguy8800, G. Thomas, Grutness, Hmains, MONGO, Mindmatrix, Pmc, RedWolf, Wandering piranga, Wikipopsicles4, 2 anonymous edits

Ravalli County, Montana *Source*: http://en.wikipedia.org/w/index.php?title=Ravalli_County%2C_Montana *Contributors*: Ahoerstemeier, Backspace, BitterrootBob, CanisRufus, Cheyinka, D6, Dakart, Floydspinky71, G. Thomas, Glacier109, GrahamHardy, Hmains, John Carter, Jruffatto, Kwamikagami, Longnskinny, Ltvine, Mike Cline, Mmm, Monegasque, Nyttend, Ram-Man, Rettetast, Sgt Pinback, Smallbones, TUF-KAT, Template namespace initialisation script, Whhalbert, WillC, Worldenc, 14 anonymous edits

Missoula County, Montana *Source*: http://en.wikipedia.org/w/index.php?title=Missoula_County%2C_Montana *Contributors*: Aericsso, Ahoerstemeier, Backspace, CanisRufus, Colonies Chris, D6, DenverLady, Floydspinky71, Funnyhat, Glacier109, GrahamHardy, Jagerz00, John Carter, Johnny Neutrino, Laobc, Lightmouse, Lincolnite, Linda Rider, Longnskinny, Ltvine, Mike Cline, Monegasque, MontanaAbe, Nyttend, PhilFree, RSM, RafikiSykes, Ram-Man, Rettetast, Robertjohnsonrj, Sgt Pinback, Smallbones, SoLando, Soglad Tomeetyou, TUF-KAT, Template namespace initialisation script, Whhalbert, Worldenc, 18 anonymous edits

Image Sources, Licenses and Contributors

file: Selway_Bitterroot Wilderness.jpg *Source*: http://en.wikipedia.org/w/index.php?title=File:Selway_Bitterroot_Wilderness.jpg *License*: unknown *Contributors*: United States Forest Service

File:Flag of the United States.svg *Source*: http://en.wikipedia.org/w/index.php?title=File:Flag_of_the_United_States.svg *License*: unknown *Contributors*: Anomie

File:Mt Banner and Thousand Island Lake.jpg *Source*: http://en.wikipedia.org/w/index.php?title=File:Mt_Banner_and_Thousand_Island_Lake.jpg *License*: unknown *Contributors*: Dcrjsr

File:Flag of Idaho.svg *Source*: http://en.wikipedia.org/w/index.php?title=File:Flag_of_Idaho.svg *License*: unknown *Contributors*: Awg1010, Denelson83, Dzordzm, Fry1989, Homo lupus, Mmaxx, Robbiegiles, Serinde, Smooth O, Svgalbertian, 2 anonymous edits

File:Seal of Idaho.svg *Source*: http://en.wikipedia.org/w/index.php?title=File:Seal_of_Idaho.svg *License*: unknown *Contributors*: Original uploader was Zscout370 at en.wikipedia

File:Idaho in United States.svg *Source*: http://en.wikipedia.org/w/index.php?title=File:Idaho_in_United_States.svg *License*: unknown *Contributors*: TUBS

Image:Speakerlink.svg *Source*: http://en.wikipedia.org/w/index.php?title=File:Speakerlink.svg *License*: unknown *Contributors*: Woodstone. Original uploader was Woodstone at en.wikipedia

File:Lakecoeurdalenebig.jpg *Source*: http://en.wikipedia.org/w/index.php?title=File:Lakecoeurdalenebig.jpg *License*: unknown *Contributors*: Original uploader was Jamidwyer at en.wikipedia

File:Idaho ned.jpg *Source*: http://en.wikipedia.org/w/index.php?title=File:Idaho_ned.jpg *License*: unknown *Contributors*: w:United States Geological Survey

File:Public-Lands-Western-US.png *Source*: http://en.wikipedia.org/w/index.php?title=File:Public-Lands-Western-US.png *License*: unknown *Contributors*: User:Northwest-historian

File:Owyhee Mountains.jpg *Source*: http://en.wikipedia.org/w/index.php?title=File:Owyhee_Mountains.jpg *License*: unknown *Contributors*: Nyttend, Robbiegiles, Shyam, Urban

File:Redfish lake.JPG *Source*: http://en.wikipedia.org/w/index.php?title=File:Redfish_lake.JPG *License*: unknown *Contributors*: User:Karthikc123

File:Idaho USA12.jpg *Source*: http://en.wikipedia.org/w/index.php?title=File:Idaho_USA12.jpg *License*: unknown *Contributors*: Bdk, Dereckson, FlickreviewR, Leoboudv, Robbiegiles, Urban

File:shoshone falls.jpg *Source*: http://en.wikipedia.org/w/index.php?title=File:Shoshone_falls.jpg *License*: unknown *Contributors*: User:Karthikc123

File:National-atlas-idaho.PNG *Source*: http://en.wikipedia.org/w/index.php?title=File:National-atlas-idaho.PNG *License*: unknown *Contributors*: 1 anonymous edits

File:Idaho population map.png *Source*: http://en.wikipedia.org/w/index.php?title=File:Idaho_population_map.png *License*: unknown *Contributors*: User:JimIrwin/Lineage_of_population_maps

File:Idaho Falls Temple.jpg *Source*: http://en.wikipedia.org/w/index.php?title=File:Idaho_Falls_Temple.jpg *License*: unknown *Contributors*: Photo by Merrill Humberg, Taken 11/25/2006. Wikipedia uploader was Twunchy at en.wikipedia

File:2007 ID Proof Rev.png *Source*: http://en.wikipedia.org/w/index.php?title=File:2007_ID_Proof_Rev.png *License*: unknown *Contributors*: Amada44, Carlomorino, CountingPine, Gabriel VanHelsing, JeffBillman, Rugby471, 3 anonymous edits

File:US ID AmericanFalls.jpg *Source*: http://en.wikipedia.org/w/index.php?title=File:US_ID_AmericanFalls.jpg *License*: unknown *Contributors*: Original uploader was Robbiegiles at en.wikipedia

File:Wheat harvest.jpg *Source*: http://en.wikipedia.org/w/index.php?title=File:Wheat_harvest.jpg *License*: unknown *Contributors*: Bdk, Inkybutton, Jurema Oliveira, Man vyi, Queryzo, 2 anonymous edits

File:Electricity Generation in Idaho.jpg *Source*: http://en.wikipedia.org/w/index.php?title=File:Electricity_Generation_in_Idaho.jpg *License*: unknown *Contributors*: User:Andrewspy007

File:Idaho license plate - passenger baseplate - 2008.jpg *Source*: http://en.wikipedia.org/w/index.php?title=File:Idaho_license_plate_-_passenger_baseplate_-_2008.jpg *License*: unknown *Contributors*: FleetCommand, Hammersoft, Presidentman, Qqqqqq, Vanished 1850

File:I-15.svg *Source*: http://en.wikipedia.org/w/index.php?title=File:I-15.svg *License*: unknown *Contributors*: Augiasstallputzer, Juliancolton, Ltljltlj, SPUI, 4 anonymous edits

File:US 95.svg *Source*: http://en.wikipedia.org/w/index.php?title=File:US_95.svg *License*: unknown *Contributors*: Bidgee, SPUI, Xnatedawgx, 2 anonymous edits

File:US 2.svg *Source*: http://en.wikipedia.org/w/index.php?title=File:US_2.svg *License*: unknown *Contributors*: SPUI, 1 anonymous edits

File:US 12.svg *Source*: http://en.wikipedia.org/w/index.php?title=File:US_12.svg *License*: unknown *Contributors*: Bidgee, SPUI, 3 anonymous edits

File:US 93.svg *Source*: http://en.wikipedia.org/w/index.php?title=File:US_93.svg *License*: unknown *Contributors*: Kanonkas, SPUI, Xnatedawgx

File:US 20.svg *Source*: http://en.wikipedia.org/w/index.php?title=File:US_20.svg *License*: unknown *Contributors*: Kanonkas, SPUI

File:US 26.svg *Source*: http://en.wikipedia.org/w/index.php?title=File:US_26.svg *License*: unknown *Contributors*: Bidgee, SPUI, 3 anonymous edits

File:US 30.svg *Source*: http://en.wikipedia.org/w/index.php?title=File:US_30.svg *License*: unknown *Contributors*: Bidgee, SPUI, 4 anonymous edits

File:I-84.svg *Source*: http://en.wikipedia.org/w/index.php?title=File:I-84.svg *License*: unknown *Contributors*: Augiasstallputzer, Ltljltlj, SPUI, 1 anonymous edits

File:I-86.svg *Source*: http://en.wikipedia.org/w/index.php?title=File:I-86.svg *License*: unknown *Contributors*: Augiasstallputzer, Ltljltlj, SPUI, 1 anonymous edits

File:I-90.svg *Source*: http://en.wikipedia.org/w/index.php?title=File:I-90.svg *License*: unknown *Contributors*: Augiasstallputzer, Ltljltlj, SPUI, 2 anonymous edits

File:I-184 (ID).svg *Source*: http://en.wikipedia.org/w/index.php?title=File:I-184_(ID).svg *License*: unknown *Contributors*: Ltljltlj

File:Boise Idaho.jpg *Source*: http://en.wikipedia.org/w/index.php?title=File:Boise_Idaho.jpg *License*: unknown *Contributors*: Emes, Infrogmation, Jareha, Schekinov Alexey Victorovich, Urban, X-Weinzar, Xnatedawgx

File:Idaho map counties.svg *Source*: http://en.wikipedia.org/w/index.php?title=File:Idaho_map_counties.svg *License*: unknown *Contributors*: User:ZooFari

File:CdA.jpg *Source*: http://en.wikipedia.org/w/index.php?title=File:CdA.jpg *License*: unknown *Contributors*: Ebcdic

File:DSCN6296 pocatellodowntown e.jpg *Source*: http://en.wikipedia.org/w/index.php?title=File:DSCN6296_pocatellodowntown_e.jpg *License*: unknown *Contributors*: User:Decumanus

File:downtownPostfalls.jpg *Source*: http://en.wikipedia.org/w/index.php?title=File:DowntownPostfalls.jpg *License*: unknown *Contributors*: Post Falls Man

File:Idaho Falls Bonneville Hotel.jpg *Source*: http://en.wikipedia.org/w/index.php?title=File:Idaho_Falls_Bonneville_Hotel.jpg *License*: unknown *Contributors*: Yassie

File:City of rocks view NPS.jpg *Source*: http://en.wikipedia.org/w/index.php?title=File:City_of_rocks_view_NPS.jpg *License*: unknown *Contributors*: Wallace Keck, Park Superintendent, City of Rocks National Reserve. Idaho

File:Scoria field at Craters of the Moon NM-750px.JPG *Source*: http://en.wikipedia.org/w/index.php?title=File:Scoria_field_at_Craters_of_the_Moon_NM-750px.JPG *License*: unknown *Contributors*: Original uploader was Mav at en.wikipedia

File:Boise2008 065.JPG *Source*: http://en.wikipedia.org/w/index.php?title=File:Boise2008_065.JPG *License*: unknown *Contributors*: User:Staplegunther

File:DSCN6179 bearlake e.jpg *Source*: http://en.wikipedia.org/w/index.php?title=File:DSCN6179_bearlake_e.jpg *License*: unknown *Contributors*: Juiced lemon, Rmhermen, Robbiegiles, Sascha Brück, 1 anonymous edits

File:Spori Building.jpg *Source*: http://en.wikipedia.org/w/index.php?title=File:Spori_Building.jpg *License*: unknown *Contributors*: Ken Lund. Original uploader was TrustTruth at en.wikipedia

File:Idaho State University.jpg *Source*: http://en.wikipedia.org/w/index.php?title=File:Idaho_State_University.jpg *License*: unknown *Contributors*: Staplegunther

File:UI-arboretum-spring-moscow-id-us.png *Source*: http://en.wikipedia.org/w/index.php?title=File:UI-arboretum-spring-moscow-id-us.png *License*: unknown *Contributors*: Original uploader was Robbiegiles at en.wikipedia

File:Smurf Turf in Bronco Stadium Idaho.JPG *Source*: http://en.wikipedia.org/w/index.php?title=File:Smurf_Turf_in_Bronco_Stadium_Idaho.JPG *License*: unknown *Contributors*: Staplegunther (talk)David Jolley. Original uploader was Staplegunther at en.wikipedia

File:Nezperceindians1895ish.jpg *Source*: http://en.wikipedia.org/w/index.php?title=File:Nezperceindians1895ish.jpg *License*: unknown *Contributors*: Dana boomer, Ealdgyth, Montanabw, Origamiemensch, Pitke, 1 anonymous edits

File:Philadelphus shrub.jpg *Source*: http://en.wikipedia.org/w/index.php?title=File:Philadelphus_shrub.jpg *License*: unknown *Contributors*: MPF, Monkeybait

file:Usa edcp relief location map.png *Source*: http://en.wikipedia.org/w/index.php?title=File:Usa_edcp_relief_location_map.png *License*: unknown *Contributors*: User:Uwe Dedering

File:Red pog.svg *Source*: http://en.wikipedia.org/w/index.php?title=File:Red_pog.svg *License*: unknown *Contributors*: Anomie

Image:BitterrootMtnsMT.jpg *Source*: http://en.wikipedia.org/w/index.php?title=File:BitterrootMtnsMT.jpg *License*: unknown *Contributors*: Original uploader was G. Thomas at en.wikipedia

Image:Deerfire high res.jpg *Source*: http://en.wikipedia.org/w/index.php?title=File:Deerfire_high_res.jpg *License*: unknown *Contributors*: John McColgan, Bureau of Land Management, Alaska Fire Service. – Alaskan Type I Incident Management Team

Image:Trappk.jpg *Source*: http://en.wikipedia.org/w/index.php?title=File:Trappk.jpg *License*: unknown *Contributors*: Lupo, MONGO, Martin H., Shanel, Thuresson, 1 anonymous edits

File:Lochsa-clearwater.jpg *Source*: http://en.wikipedia.org/w/index.php?title=File:Lochsa-clearwater.jpg *License*: unknown *Contributors*: Original uploader was Robbiegiles at en.wikipedia

File:Rainy Lake in Lolo National Forest.jpg *Source*: http://en.wikipedia.org/w/index.php?title=File:Rainy_Lake_in_Lolo_National_Forest.jpg *License*: unknown *Contributors*: Lupo, MONGO, Martin H., Urban

file: River_of_No_Return_Wilderness.jpg *Source*: http://en.wikipedia.org/w/index.php?title=File:River_of_No_Return_Wilderness.jpg *License*: unknown *Contributors*: Original uploader was Hike395 at en.wikipedia

Image:FCRONR Wildernes Map.jpg *Source*: http://en.wikipedia.org/w/index.php?title=File:FCRONR_Wildernes_Map.jpg *License*: unknown *Contributors*: USFS

File:Crooked Creek in Gospel Hump Wilderness.jpg *Source*: http://en.wikipedia.org/w/index.php?title=File:Crooked_Creek_in_Gospel_Hump_Wilderness.jpg *License*: unknown *Contributors*: Howcheng, Jsayre64, Lupo, MONGO, Martin H., Urban, 1 anonymous edits

Image: Lochsaheadwaters.jpg *Source*: http://en.wikipedia.org/w/index.php?title=File:Lochsaheadwaters.jpg *License*: unknown *Contributors*: User:Finetooth

Image: Selway-river-id.png *Source*: http://en.wikipedia.org/w/index.php?title=File:Selway-river-id.png *License*: unknown *Contributors*: Robbiegiles

Image: Idaho Locator Map.PNG *Source*: http://en.wikipedia.org/w/index.php?title=File:Idaho_Locator_Map.PNG *License*: unknown *Contributors*: US Census, Ruhrfisch

Image:WaltonCreekLochsa.jpg *Source*: http://en.wikipedia.org/w/index.php?title=File:WaltonCreekLochsa.jpg *License*: unknown *Contributors*: User:Finetooth

Image:LochsaRiverKayakByPhilKonstantin.jpg *Source*: http://en.wikipedia.org/w/index.php?title=File:LochsaRiverKayakByPhilKonstantin.jpg *License*: unknown *Contributors*: User:Philkon

Image:Selway-river-id.png *Source*: http://en.wikipedia.org/w/index.php?title=File:Selway-river-id.png *License*: unknown *Contributors*: Robbiegiles

Image:Selway-river-1.jpg *Source*: http://en.wikipedia.org/w/index.php?title=File:Selway-river-1.jpg *License*: unknown *Contributors*: Robbiegiles

Image: ClearwaterRiver_nearGreerID.jpg *Source*: http://en.wikipedia.org/w/index.php?title=File:ClearwaterRiver_nearGreerID.jpg *License*: unknown *Contributors*: Idaho Department of Commerce and Labor

Image: Clearwateridrivermap.png *Source*: http://en.wikipedia.org/w/index.php?title=File:Clearwateridrivermap.png *License*: unknown *Contributors*: User:Kmusser

Image:03-37-14.jpg *Source*: http://en.wikipedia.org/w/index.php?title=File:03-37-14.jpg *License*: unknown *Contributors*: User:Dsdugan

Image:TrapperPeakMT.jpg *Source*: http://en.wikipedia.org/w/index.php?title=File:TrapperPeakMT.jpg *License*: unknown *Contributors*: G. Thomas

file:USA Montana location map.svg *Source*: http://en.wikipedia.org/w/index.php?title=File:USA_Montana_location_map.svg *License*: unknown *Contributors*: User:Alexrk2

File:TrapperPeakMTfromRt93.jpg *Source*: http://en.wikipedia.org/w/index.php?title=File:TrapperPeakMTfromRt93.jpg *License*: unknown *Contributors*: Wandering piranga

File:Map of Montana highlighting Ravalli County.svg *Source*: http://en.wikipedia.org/w/index.php?title=File:Map_of_Montana_highlighting_Ravalli_County.svg *License*: unknown *Contributors*: User:Dbenbenn

File:Map of USA MT.svg *Source*: http://en.wikipedia.org/w/index.php?title=File:Map_of_USA_MT.svg *License*: unknown *Contributors*: Abnormaal, Hogweard, Huebi, Lokal Profil, Lupo, Mattbuck, Petr Dlouhý, 2 anonymous edits

Image:US 93.svg *Source*: http://en.wikipedia.org/w/index.php?title=File:US_93.svg *License*: unknown *Contributors*: Kanonkas, SPUI, Xnatedawgx

Image:MT-38.svg *Source*: http://en.wikipedia.org/w/index.php?title=File:MT-38.svg *License*: unknown *Contributors*: User:Master_son

File:Map of Montana highlighting Missoula County.svg *Source*: http://en.wikipedia.org/w/index.php?title=File:Map_of_Montana_highlighting_Missoula_County.svg *License*: unknown *Contributors*: User:Dbenbenn

Image:I-90.svg *Source*: http://en.wikipedia.org/w/index.php?title=File:I-90.svg *License*: unknown *Contributors*: Augiasstallputzer, Ltljltlj, SPUI, 2 anonymous edits

Image:US 12.svg *Source*: http://en.wikipedia.org/w/index.php?title=File:US_12.svg *License*: unknown *Contributors*: Bidgee, SPUI, 3 anonymous edits

Image:MT-83.svg *Source*: http://en.wikipedia.org/w/index.php?title=File:MT-83.svg *License*: unknown *Contributors*: User:Master_son

Image:MT-200.svg *Source*: http://en.wikipedia.org/w/index.php?title=File:MT-200.svg *License*: unknown *Contributors*: Kanonkas, PHenry

MIX
Papier aus verantwortungsvollen Quellen
Paper from responsible sources
FSC® C105338

Printed by Books on Demand GmbH, Norderstedt / Germany